IMPACT JUPITER

The Crash of Comet Shoemaker–Levy 9

IMPACT JUPITER
The Crash of Comet Shoemaker–Levy 9

DAVID H. LEVY
Co-Discoverer of the Comet

PLENUM PRESS • NEW YORK AND LONDON

Library of Congress Cataloging-in-Publication Data

Levy, David H., 1948-
    Impact Jupiter : the crash of comet Shoemaker-Levy 9 / David H.
  Levy, co-discoverer of the comet.
      p.   cm.
    Includes bibliographical references and index.
    ISBN 0-306-45088-7
    1. Shoemaker-Levy 9 comet--Impact. 2. Jupiter (Planet)--Impact.
  I. Title.
  QB723.S26L48   1995
  523.6'4--dc20   95-33339
                                                             CIP

Cover and chapter-opening page illustration: One at a time, the fragments of Comet Shoemaker–Levy 9 collide with Jupiter. Painting by Kim P. Poor.

ISBN 0-306-45088-7

For Gene and Carolyn, in friendship and respect.
May all your nights be clear and
all your comets bright.

The big door of Palomar's 60-inch telescope building slammed behind us as we walked out into the gathering storm. This was the cold, wet winter of 1993: dense clouds, thick fog, high winds, and snowflakes. Gene Shoemaker and I had just deposited two boxes of film in an oven, where baking them for six hours would increase their sensitivity to light. Fat chance we'd ever get to use these films this run! All we wanted to do now was drive the half mile to the warmth of our own little observatory building.

Four of us were at the mountaintop that month. Eugene and Carolyn Shoemaker have been hunting comets and asteroids from Palomar seven nights each month for years. I joined them in August 1989. This run, French astronomer Philippe Bendjoya had come along to find out what observing with a wide-field camera means.

Philippe was a quick study: Even in famously sunny southern California, the weather can be as poor as it often is in his native France. And on this night the weather was pretty awful; there was little chance that we'd be able to open the telescope. Moreover, we heard that the slow-moving storm would take several days to pass through. Maybe we would not get to open the telescope any more at all. That was a sobering thought.

We drove down the hill and back along the road to the 18-inch Schmidt telescope. It's an instrument with a long and rich history. The first telescope built on Palomar Mountain, and opened in 1936, it was designed partly by Russell W. Porter, an amateur astronomer who, it is said, brought the art of telescope making "out of the woods and into the basement." Thanks to Porter, small and inexpensive telescopes sprang up all over the world. Porter's fame spread in 1925, when an article in *Scientific American* showed how easy he had made the art of grinding a mirror and making a telescope. On the west coast, George Ellery Hale, hearing of Porter's work, invited him to come to California and help design Palomar's new 200-inch telescope. For years it would be the largest telescope in the world. But the big scope would be many years in coming. Responding to a need to begin a supernova search on Palomar Mountain, Porter designed a beautiful wide-field camera, a Schmidt telescope through which a single exposure would cover a large expanse of sky. By 1936, the telescope was finished and saw first light of the stars. Its discoveries are legion: 47 comets, perhaps a hundred supernovae, many asteroids, and one fictitious globule that later swallowed the Earth. Spread across the mountaintop are a 200-inch telescope, a 60-inch, a 48-inch, and our 18-inch.

On the way back to the 18-inch, Gene reminisced a bit about the weather he'd seen here in the nearly 20 years he had been observing here. "Some winters are just great," he said. "This isn't one of them."

That was an understatement. But even though the weather was bad this season, Gene had a lot to be thankful for. His years at Palomar had been rewarding and productive. With his wife Carolyn, he had surveyed much of the sky more than 40 times in search of asteroids and comets.

He is interested in these objects because someday, some of them will collide with the planets, including, especially, the Earth.

Gene's whole career has been centered around impacts. In his mid-1950s Ph.D. thesis he proved that the big crater east of Flagstaff, Arizona, was the result of an asteroid that hit the Earth some 50,000 years ago. During the 1960s and 1970s, he studied the lunar impact craters revealed on the Moon by the many manned and unmanned missions sent there. And for the last 20 years, his search had yielded a bonanza of comets and asteroids.

Seeing an actual impact take place was another matter entirely. A big one doesn't happen on Earth very often, and in all of history no one had ever confirmed seeing an impact on another planet.

But we could always dream. What happened, I asked him, when the object apparently responsible for the extinction of the dinosaurs hit the Earth 65 million years ago? The picture he painted was frightening. First, there was a gigantic fireball brighter than the Sun as the comet plunged to its death, not with a whimper, but a bang. One casualty was the ozone layer, which temporarily vanished. Seconds after the big comet first encountered Earth's upper atmosphere, it carved out a crater—now buried—200 kilometers wide and 25 kilometers deep. All that debris shot up into the sky and came back again, all over the Earth. No place would have been spared a hit of at least a tiny particle.

Reacting to this incredible bombardment, the air temperature rose quickly until, for more than two hours, the worldwide temperature reached that of an oven set to broiling. The sky glowed like an electric heater. Ground fires flared everywhere. Then the temperature started to drop, and drop, and drop. A thick cloud of dust blackened the world, setting off a several-month period without sunlight. Rains poisoned with sulfuric and nitric acid added to the misery.

With blow after blow to the biosphere, Gene guessed, most large land-roving dinosaurs probably died within weeks. Other creatures took longer; those who survived one disaster would perish in the next one. Slowly, the great cloud dissipated, and temperatures began to rise again, this time due to a greenhouse effect that lasted for centuries or millennia. Overall, perhaps 70 percent of all the species of life died during the siege, and in North America at least, about half of the species of flowering plants.

But not everybody. Some of the hardier representatives of many species, including the ones equipped to hibernate, made it through the impact winter. Enough small mammals survived that, when the biosphere finally started to recover, they began to proliferate and flourish.

"Would it be correct to say," I inquired of Gene, "that this impact was good for us?" Impacts clear the decks for new forms of life, Gene was trying to say. The fossil record shows that after major impacts, there is a burst of speciation. New life forms fill the niches that the old ones leave behind. If there were no impacts, the thrust of evolution might have slowed down, and today there would be a different set of species inhabiting the Earth.

We were now back at the 18-inch telescope. The storm was getting worse, but somehow I felt better knowing that in the past few minutes we had conjured up a world that was pure hell, a place that made a stormy winter at Palomar look like a picnic.

Acknowledgments

Many people were involved in the saga of Shoemaker–Levy 9. Jim Scotti, Steve Edberg, Don Yeomans, and Paul Chodas were very helpful. The following provided useful information: Ron Baalke, Patrick Vanouplines, James DeYoung, Brent Archinal, Lonny Baker, Jean Mueller, Hal Weaver, Thomas Hockey, Clyde Tombaugh, Patsy Tombaugh, James Klavetter, Reta Beebe, Heidi Hammel, Michael Liu, Elizabeth Roettger, Anne Sprague, Don Hunten, Walter Wild, Peter Jedicke, and Tom Hill.

Sandy Sheehy, Gene and Carolyn Shoemaker, Wendee Wallach, Peter Jedicke, Brian Marsden, Steve Edberg, Leo Enright, and Daniel Green spent much of their valuable time reading the manuscript. It has been a real pleasure to work with Linda Regan, Kenn Schubach, and their colleagues at Plenum. All these people helped make this a more accurate and interesting book, and I am grateful for their assistance.

Contents

Of Comets, Planets, and People

The Oort cloud is a long way from Earth. In fact it isn't a cloud at all but a spherical halo that surrounds the Sun way beyond the orbits of Neptune and Pluto. It is possible that Comet Shoemaker–Levy 9 was once a member of this sphere. It is also possible that the comet was once a lot closer to the Sun, either in a ribbon of comets past Neptune and Pluto that we call the Kuiper Belt or even closer to Earth mixed in among the asteroids.

Some time in the distant past, possibly a hundred million years ago or more, some gravitational tug caused the comet to leave its comfortable home and take a different path. After a long series of loops about the Sun, it passed near Jupiter and began a stately dance, passing close to the planet many times. But one of these passages was a critical one. Hardly more than 10 kilometers (about 6 miles) in diameter, the little comet had what astronomers call a low-velocity encounter with Jupiter around 1929. Moving at much the same speed as Jupiter, the comet curved sharply and took on an entirely new path. Where once the comet swung round the Sun, now it recognized mighty Jupiter as the object about which it would travel.

But the new orbit was never a stable one, like those of the 16 Jovian moons that have been circling Jupiter for eons. In chaotic style, the interloper changed its path almost every time it reached its closest distance from Jupiter, or its perijove, every orbit a different one. Had the comet had any inhabitants, they would have been treated to a string of different and dramatic views of Jupiter.

By the start of 1992, the comet was heading for its closest encounter yet with the giant planet. With each passing day the planet loomed larger on the comet's horizon, and by the first of July the comet seemed to be making a kamikaze dive into Jupiter's clouds. Day after day saw Jupiter dominate more and more of the sky.

On July 7 the comet missed Jupiter by the perilously close distance of 20,000 kilometers, or 13,000 miles. Flying over huge storm systems at some 50 kilometers per second, the comet seemed as if it would zoom by the whole planet in less than an hour. It didn't make it. Jupiter's powerful gravity tore at the comet. Like a giant hand reaching up and ripping the comet apart, Jupiter's gravity pulled on the closest part harder than it pulled on the most distant. As the comet started to stretch out like a noodle, with a shudder it simply came unglued. I don't remember what I was doing on July 7, 1992, but I am sure that my day was not as stressful as that comet's.

Having almost hit Jupiter, the comet came out of its dive transformed. No longer one comet, it was now maybe 20. Huge amounts of dust were escaping as each nucleus formed its own cloud of dust, called a coma, and a tail. The rest of the dust was spreading out on either side of the string of fragments, and the whole complex was getting brighter by the day.

Almost 600 million miles away, no one on Earth was aware of the remarkable drama unfolding in the sky over Jupiter. The reason: Jupiter was on the opposite side of the Sun from Earth, so no one was watching. By the end of autumn, Jupiter was rising several hours before dawn, and anyone with a large telescope should have detected the comet, now a considerable distance from Jupiter. But the comet was too faint for amateur astronomers searching visually through small telescopes, and it was not yet near opposition, the region of the sky opposite the Sun where the big search cameras concentrate.

JUPITER WATCH

On a clear, mild night in August, 1960, my parents and I set up my new telescope. Its 3½-inch mirror was tiny compared to the mirrors I had read about, but to me this instrument was the best in the world. Echo was my first telescope, and on this night, it was going to give me my first look at the heavens. Not knowing the identity of any of the stars, I decided to turn the telescope to the brightest one, which hung in the southern sky between two trees. I will never forget that first look, and the feeling I had when I saw that the bright star was not a star at all but Jupiter, complete with a dark belt and four moons.

Almost 33 years later, on a night in January 1993, high clouds obscured the sky above my home and observing site in the desert southeast of Tucson, Arizona. There didn't seem to be any point in going out to observe at all, but I felt in the mood for a little viewing. The cirrus was so thick, however, that Jupiter appeared to be the only thing in the whole sky. I turned the telescope toward Jupiter, put in a high-power eyepiece, and observed the closest of the solar system's planetary giants for a while.

Even on a mostly cloudy night like this, Jupiter is well worth watching. Its dense atmosphere is dominated by two dark belts, one on either side of the equator. Despite the clouds, the view was a steady one, and I could make out detail within each belt, as well as some thin dark bands, called festoons, that fall away from the belts into the bright equatorial zone between them. There was also a column of dark material that stretched across the zone from one belt to the other.

I keep a telephone out in the observatory, and since I knew that my friend Jim Scotti was that very night at nearby Kitt Peak National Observatory, I decided to call him. High above the desert floor in the Quinlan Mountains southwest of Tucson, Jim was at the controls of a 36-inch reflector telescope on which a modern electronic detector called a CCD (or charge-coupled device) took the place of an eyepiece. But for all his fancy gadgetry, Jim wasn't able to do any more than I was. He was sitting under the same cloud.

At least I had Jupiter to marvel at. What a piece of work this planet is! Some 11 times the diameter of the Earth, it spins its huge bulk around in 9 hours 55 minutes. So fast is the planet's rotation that Jupiter's north and south poles are considerably flattened, giving the planet an oval look. Whenever I draw the features on Jupiter, I have to get the basic outline done very quickly, within 10 minutes; otherwise the planet moves its eastern features out of sight.

Even though the sky was pretty cloudy, I could still make out Jupiter's four bright moons: Io, Europa, Ganymede, and Callisto. I remember observing these moons back in January 1963, with the same small telescope I had used to take my first look at Jupiter three years earlier. But this was not just an observation, it was a project. I was watching the slow workings of a clock in space; each night the moons had different positions. Io was everywhere, racing around Jupiter each day, its position was a complete surprise on the nights it wasn't hidden behind or in front of the big planet. Europa moved more slowly, but Ganymede and Callisto, the two outermost moons, really impressed me. They marched in stately fashion slowly around Jupiter, patiently ticking off with each orbit the slow march of time in the solar system. For a 14-year-old rushing through life, it was a good lesson in patience.

In the years since Galileo discovered these moons in 1610, few new facts or ideas were announced regarding Jupiter. In 1672, Giovanni Cassini described what appears to be the first observation of the famous Great Red Spot.[1] In 1892, the American astronomer Edward Emerson Barnard found a new moon, now called

Amalthea. In 1979, two Voyager spacecraft took exquisite pictures as they encountered Jupiter at close range. The total number of moons was now up to 16. Io had several volcanoes actually erupting as the spacecraft passed by; one has a plume of sulfur dioxide and sulfur hanging over the moon's yellow surface. Europa was a moon-wide skating rink. The largest moons, Ganymede and Callisto, were scarred by the crater souvenirs of many comets.

One small mystery remained after the Voyagers left Callisto and planetary scientists examined their last pictures. There were 16 chains of craters scattered across the side of Callisto that always faces Jupiter. Planetary geologists were baffled by these chains. Could they be evidence of secondary crashes made by large pieces that were gouged out after a big comet hit? But this scenario would require that some large crater be nearby, and none existed for any of the 16 crater chains. Shrugging their shoulders, investigators resigned themselves to wait for something to help explain these mysterious marks on the face of Callisto.

Of course, I could see none of these intricate details on that cloudy January night. As I described what I was seeing on Jupiter to Jim Scotti, he said he wished he could take a look as well. The festoon that dropped out of the southern equatorial belt had moved eastward along with the planet's rapid rotation. In just five hours it would lie clear across on the other side of the planet.

CLOSER TO HOME

That night, Jim Scotti spoke of the asteroid and comet work he was up to. In the past couple of years the Spacewatch program had located a host of small asteroids that were on near-collision courses with the Earth. Some as small as 10 meters across, these objects would normally be too faint to spot even through the mighty electronic eye of the 36-inch diameter Spacewatch camera. But if an asteroid gets close enough, the system might spot it moving at maybe 7 degrees of arc, or about 14 moon diameters, each day as it sprints across the sky. These things are so small that even if one were to crash into the Earth, it would do no major damage. It would appear only as a very bright fireball in the sky, possibly depositing a few football-sized chunks of rock on the ground.

Jim Scotti has been with the Spacewatch project since 1983, when it was just a fledgling operation. Inspired by Tom Gehrels, a top asteroid specialist, Spacewatch aims to discover asteroids on courses that intersect the Earth's orbit. "Find them before they find us!" the Spacewatch motto cries. Since April 1984, when they found their first new asteroid, Spacewatch has located some of the most interesting asteroids. One of them, labeled 1991 VG, was orbiting the Sun in a path that matches the orbit of the Earth closely enough to suggest that the body is an artificial one, perhaps the upper-stage rocket of a space probe launched years ago and sent into an orbit around the Sun.

Although it wasn't a part of the original Spacewatch mission, Jim has used the telescope to check on comets that were expected to return in their courses about the Sun. In July 1985 he "recovered" his first, Periodic Comet Whipple. The comet was near perihelion, or the closest point to the Sun, of its 8½-year orbit. By the end of 1994 Jim had recovered 48 comets. So often does this discovery–recovery process take place that our phone conversations begin with a predictable routine: I ask him whether he has recovered any comets lately. He answers by telling me what he has seen and what he hopes to see in the coming days and weeks. "Now David," he always adds, "have you discovered any comets recently?"

But this Friday night we forgot the routine. Unfortunately, the cloudy sky hadn't helped alleviate the long dry spell since my last visual comet find in June 1991. And it had been almost a year since the Shoemakers and I found a new one through our photographic program. Over the phone this Friday night, I asked Jim if he had succeeded in getting an image of a comet I had discovered 18 months earlier. (Periodic Comet Levy was going out on a long 50-year loop away from the Sun, leaving behind the unanswered question of whether it just might be the same comet as one that passed close to the Earth in 1499. The more accurate positions we can get of this comet, the greater the chance we will learn if the two discoveries really were apparitions of the same comet.) Yes, Scotti answered, using his telescope and CCD, he did catch the comet. Once a bright feathered shuttlecock visible through small telescopes, it was now reduced to a barely detectable faint smudge.

I first met Jim Scotti when he was a mere sophomore at the University of Arizona. Bright-eyed with enthusiasm for the sky, he worked part-time at the university's Flandrau Planetarium. For some years we observed together on clear and moonless nights. I remember one time we made a big effort to see Periodic Comet Brooks 2. Coming in on its thirteenth visit since William Brooks, an amateur comet sleuth from Phelps, New York, discovered it on July 7, 1889, the comet has an eventful history. On August 1, Barnard noticed that it had two faint companion comets; soon afterward several other miniatures turned up, each with its own coma and tail. Like a bear walking with her cubs, the comet wandered on its path through the sky.

How did all this happen? During a two-day period in 1886, the comet passed much closer to Jupiter than Io's distance: It passed within what's called the planet's Roche limit. A nineteenth-century French mathematician, Edouard Roche proposed that if a body that was loosely held together to start with passed closer than a certain distance—called the Roche limit—from a planet, tidal forces would tear that body apart. That's how Saturn's rings likely formed. And that is how Periodic Comet Brooks 2 fell apart.

But Brooks' baby comets apparently didn't survive. Two disappeared within a few days, and all had vanished by the end of November. Since then, Brooks 2 has not ventured that close to Jupiter. On August 17, 1987, Jim Scotti and I tried hard, using my 16-inch reflector and a high-power eyepiece, to see just one very faint comet. Peering intently through the telescope, we could barely see the comet's fuzzy image. Its prodigal children were nowhere to be seen. The comet was so faint, in fact, that each of us drew its position among the background of stars in the sky, so that afterward, we could compare our sketches and be sure that we had seen the same thing. A later check of a star atlas showed that our comet was exactly in the right spot. Next came estimating the comet's magnitude, a tricky game of comparing the comet's brightness with the known magnitudes of stars.

IF IT MOVES OR EXPLODES, MARSDEN WANTS TO KNOW ABOUT IT

Our observation completed, I turned on my computer and modem to report what we'd seen to Brian Marsden, Director of the International Astronomical Union's Central Bureau for Astronomical Telegrams (CBAT). The Bureau is based at the Smithsonian Astrophysical Observatory in Cambridge, Massachusetts. A simple magnitude estimate is far from the most important type of observation that Marsden receives. He would publish it on his postcard-sized *IAU Circulars* only if he had room left over after reporting discoveries and other vital information about comets, interesting asteroids, or distant detonating suns called supernovae. Sort of a traffic cop of the sky, Marsden keeps track of where all the sky's transient objects are and what is happening to them. We e-mailed him the results of our difficult visual observation.

Jim and I had fun observing Brooks 2 that night. The radio I keep in my backyard observing shed was tuned to Tucson's oldies station, and Beatles music wafting in the background helped set our mood. After the two hours it took to find Brooks 2 we were pleased with our success. Even our message to Marsden was playful in spots, spiked with birthday wishes and gossip on what everybody else was up to.

Born August 5, 1937, Brian Marsden was, I suspect, calculating orbits before he could talk. After receiving his doctorate from Yale, he joined the CBAT on the same day that astronomers realized that Comet Ikeya-Seki, found by Kaoru Ikeya, a piano factory worker, and Tsutomu Seki, a guitar instructor, would pass within 300,000 kilometers of the surface of the Sun to become one of the century's brightest comets. By the start of 1968 Marsden was CBAT's director, a position he has served with distinction.

One of the good things about Marsden is that he clearly understands the different needs of the visual observer, who is usually an amateur astronomer, and the photographic or electronic observer, who is more often a professional astronomer employed at a university or an observatory. While he expects accurate work from everyone, Marsden is inclined to be supportive of the legions of visual observers armed with small telescopes, a lot of patience, and usually no funds—apart from what they scrape together from their day jobs. With its different agenda, the professional side of astronomy funnels a large volume of observations through Marsden.

To handle this load, the CBAT has an associate director, Daniel Green, whose name undersigns most of the IAU Circulars. As a high school student he launched a journal called *The Comet*; he still publishes it as the *International Comet Quarterly*.

As the people in charge of the world's clearinghouse for astronomical discoveries, Marsden and Green receive many reports from astronomers, mostly amateur but some professional, about the discoveries of new comets. Many reports they receive are false alarms. Because they also look fuzzy, galaxies or clusters of stars can be easily mistaken for comets; so, in the case of photographic discoveries, can a defect on a film or a reflection of some sort. Bright stars produce ghost images, which are often reported as comets, especially by the uninitiated. That's why Marsden announces no new comet until it has been seen on at least two nights and confirmed by an experienced observer.

Over the years Jim Scotti has confirmed several of my comets. In one case he drove down to my discovery site and observed Comet Levy 1987y through my own telescope, but most often he spots them with the Spacewatch camera. One of these discoveries was quite a disappointment for him. Progressing through an ordinary scan for moving objects, the Spacewatch computer system automatically detected the fuzzy trail of a faint comet. For a few moments Scotti's heart leapt as he anticipated a new comet notch on Spacewatch's impressive record of finds. But soon he learned that the comet in his field of view was one that the Shoemakers and I had picked up just a few days earlier. Now called Periodic Comet Shoemaker–Levy 5, this comet easily could have escaped us. Thanks to the dust in the upper atmosphere left by the eruption of Mount Pinatubo in the Philippines, it was hard to spot anything faint. We did it by attaching a yellow filter to the filmholder; the filter lessened the dust's contribution to the brightness of the sky.

As we were winding down our Friday night phone chat, Jim asked when I would be going back to Palomar. Just two weeks from now, I said. Jim would be searching out little rocks that could hit the Earth, and at the same time I'd be looking through a telescope at mighty Jupiter, but neither of us realized how closely our efforts would be connected. Not 3 degrees away from where my telescope pointed that Friday night, a yet undiscovered comet was hurtling through space on its way to a destiny with Jupiter.

A New Comet

Observing the Shoemaker–Levy way is a mixture of astronomy, geology, long hours at a telescope, and good humor. We try to find comets and asteroids, objects capable of making craters here and on other worlds. Each month we drive from Arizona to Palomar Mountain in southern California, north of San Diego. We use Palomar's 18-inch-diameter Schmidt telescope—actually a special camera astronomers use to photograph large segments of the sky—for seven nights in the hope that at least four of these will be clear.

Before we start to observe, Gene, Carolyn, and I prepare a large number of photographic films. To do this, Gene Shoemaker first uses a puncher that cuts a 6-inch-diameter circle out of an 8 × 10-inch sheet of film. Then we bake these films for six hours in an oven while a mixture of nitrogen and hydrogen gas blows over them. This process makes the film far more capable of picking up the faint objects we seek. The technical name for it is hypersensitization, or just hypering. We call it baking the cookies.

On clear nights we take as many as 50 or 60 exposures of

A peaceful September afternoon at Palomar Observatory. The 18-inch dome is at left, its closed shutters pointing south. At right is the 60-inch telescope dome, inside which we hyper our photographic films. Photograph by Jean Mueller.

various regions of the sky. The exposures are divided into several sets of four or five fields each. After we photograph the regions of sky, we repeat the exposures; in other words, we cover each field twice, with some 45 minutes to an hour separating the exposures. Any object that moves during that time reveals itself as a tiny blip that appears to "float" atop an otherwise static background of stars—at least to a pair of eyes practiced in the art of looking in stereo. Carolyn has one of the best.

But the year 1993 was giving our team little chance to find many objects, new or old. In January we had only one clear night out of seven, and February's observing run (a term we use to describe our periods at the telescope) offered us just one clear hour. We usually play music in the dome while we are observing, but on this frigid February night the sky began to clear just as President Clinton was about to begin his address to Congress. So as Clinton discussed the deficit and other matters of state, we took as many films as we could. Even though his address lasted an hour, we thought it was too short. As he began winding down, thick clouds gathered again in the west. "No, no," we cried, hoping he would hear us, "keep talking!" But just as he completed his remarks, the sky closed in again. That was all the observing we did that month.

With plenty of film already hypered from January and February, we arrived for our March run full of hope. The first night looked very clear indeed. A front had just passed through, and the low clouds that hung over San Diego and Los Angeles made our sky as dark as if we were on the Moon. It was the kind of night that astronomers dream of.

At least it was until Gene looked at the first two films he processed: they were absolutely, totally black. Apparently some curious soul had opened our box of precious hypered films some time in the cloudy weeks we were away from the telescope. So here we were, the first clear night in months, with no film ready to use.

The unexposed films were neatly stacked in the bottom of the box. After a few minutes, Gene decided to process one of the films

Gene, Carolyn, and the author at the 18-inch telescope at Palomar Mountain Observatory. Photo by Terence Dickinson.

Carolyn Shoemaker is about to scan a pair of films taken with the 18-inch at Palomar. Designed by Gene Shoemaker, the stereomicroscope allows scanning of two exposures at the same time. A moving comet or asteroid appears to "float" atop the starry background. It is an elegant way to search for new members of our solar system. Photo by Terence Dickinson.

farther down in the box. Perhaps, he reasoned, the top films had protected the lower ones. The result of this test was encouraging; the film was black around its edges, but the middle was not too bad. We were back in business.

On this particular observing run, a young astronomer from France was accompanying us. A specialist in families of asteroids, Philippe Bendjoya was hoping to begin an observing program using the Cerga Schmidt telescope near Nice and wanted to gain some experience using similar equipment. Having learned a little French while growing up in Montreal, I enjoyed conversing with Philippe during these cloudy nights.

By the start of the second night, March 23, we were back to our normal *modus operandi*, with a good amount of freshly hypered film. But as we finished the first set, it started to cloud over once again. Thin cirrus clouds moved across the sky in advance of an approaching storm. Once again we had to stop observing.

As we stood outside the observatory gazing forlornly at the sky, I suggested that maybe it wasn't bad enough to stop. Gene and Carolyn laughed. "It's the Levy enthusiasm!" Gene said. I just didn't want to lose yet another night to clouds. As we continued to gaze up silently, I thought that the sky might not be that bad. But Gene laughed again.

"David, every time we slap a film into that telescope, it costs four dollars. The bottom line is, we just can't afford to be taking films on a night like this." We stood there a while longer. I thought of one last angle. "What about those light-struck films," I asked. "They're already wasted. Don't we have a few left?"

"Yes, we do indeed. We've got about a dozen!" Gene said. "Let's get to it!"

We went inside, and Gene loaded a film into the circular metal film holder. Upstairs in the observatory, I swung the telescope down and opened its film door. Gene presented me with a film holder. I loaded it into the telescope, a process that was more magic than anything else. I couldn't see inside. With one hand on the film holder and the other on the telescope's locking mechanism, I felt the film holder's way inside until I could tell by feel that it was loaded correctly. Usually this takes only a few seconds. Gene knew the procedure was finished when I closed the telescope door, presented him with the film holder's cover, and turned on the telescope's position lights. Gene then read me two sets of numbers pinpointing the section that determined where we were going to shoot: the field's right ascension (a celestial projection of earthly longitude) and the declination (which corresponds to latitude.)

When the telescope arrived at the correct position, I could barely see the faint star I would fix on to help me guide the telescope for the next eight minutes. There was a bright glow in the eyepiece! I looked up and found the source of the glow. "Gene," I complained, "Jupiter's gonna be right in this field!"

Gene looked at his program list and agreed. "But this is the next field," he answered. "We're there. Let's just do it anyway."

"Okay!" I called as I centered the guide star on the illumi-
nated grid in the eyepiece. "Guide star is ready." Looking at our
universal time clock, Gene began his countdown. "Five, four,
three, two, one, OPEN!" With a squeak the shutters opened at the
top of the telescope, and the exposure was on.

Guiding an eight-minute exposure is a thoroughly pleasant
task, the object of which is to keep the eyepiece grid centered on
the star. While Gene goes downstairs to load the other film holder,
I can relax, enjoy the sky, and listen to music playing. The music
that gets played during observing varies widely, from Mozart's
Jupiter symphony to Mary-Chapin Carpenter's "When Halley
Came to Jackson." Just about anything goes, except, the Shoemakers entreat, the Beatles.

I looked up toward the sky for a second. Even though there
were cirrus clouds up there, the cloud cover was light, and I
thought this could be a successful exposure. What I wasn't so sure
about was whether the sky would remain this good for at least 45
minutes, the interval we needed before taking the second photograph. After Gene or Carolyn developed the films, Carolyn examined the two time-spaced shots of an identical field using a stereomicroscope. Many years ago, Heidelberg asteroid hunter Max
Wolf had thought of using a stereomicroscope. But, the story goes,
since one of his eyes was not good, he preferred instead to use a
blink comparator, which alternates the view between two pictures, showing only one at a time. In 1930 Clyde Tombaugh used a
blink comparator to discover the planet Pluto.

Finally, more than a decade ago, Gene devised a stereomicroscope, based on his experience as a geologist. Using stereomicro-
scopes has allowed him to examine two pictures of the same
region of Earth taken from very slightly different angles; the resulting pair provide a three-dimensional impression of depth.

As the exposure neared its end, Gene opened the door and
climbed the dozen steps to the observing floor. He gave me another countdown, and I closed the shutter. Changing the film,
from shutter closing to shutter opening, is a procedure I try to
do as quickly as possible; every second lost during this time is a

second when the film is not collecting photons. I try to complete the whole procedure in about 90 seconds.

This set included four fields, but by the middle of the third exposure the clouds were thickening. We gave up on the last exposure and worried about whether we'd get to play the bottom half of this photographic inning. Would the sky clear enough so that we could take these three exposures over again? For some time we waited, checking the sky often. Finally a small break in the clouds seemed headed our way. I loaded the film, set the telescope on the guide star, and waited until the field was more or less in the clear. Time since the first exposure started: 1 hour, 47 minutes. I barely managed to finish the eight-minute exposure before the clouds came in yet again. By and by I finished the other two. Aside from some exposures that night and early the following night, that pretty well marked the end of our observing run. Meanwhile, Gene developed the films.

Like all astronomical films, these emerged as negatives, with bright and dark reversed. After Gene took the films out of the processing chemicals, he was puzzled by the appearance of a big black splotch on two of them—maybe the light leaks were worse than he had thought. For a moment he glared at the films. But this splotch was different: It had exactly the same size and shape in each film. Gene laughed as he figured it out. This time the blobs belonged there. They were the glow from the brilliant planet Jupiter, right on our fields.

THIS LOOKS LIKE A SQUASHED COMET!

By the afternoon of March 25, the wind and snow had arrived. Remembering our nearly disastrous experience from the first night, we didn't want to be without new film if the sky cleared, so that afternoon Gene cut some new films, and he and I made the trip to the hypering oven, which is located in the 60-inch telescope dome about a mile away from our telescope. We returned to our dome feeling discouraged.

The time was 4 p.m. Once we'd scanned the films from our good first night and the few good ones from our second night, Carolyn inserted the two Jupiter films I had exposed that second night under the stereomicroscope. They did not look promising. The separation in time was more than twice as long as we would have liked. Asteroids that would normally float comfortably above the starry background would now be so far apart that it would be hard to identify them. However, the Shoemakers considered Jupiter fields to bring good luck. They had discovered a comet on a Jupiter field before—it was named Shoemaker–Holt—and was the last comet designated in 1987.

It had been almost a year since Carolyn had found a comet. Looking somewhat dejectedly at the films, she announced, "You know, I used to be a person who found comets." As Carolyn scanned, I pulled out my small laptop and worked on *The Quest for Comets*, the book I was writing. The chapter described how comets, striking the primordial Earth, deposited the original organic materials that formed the basis of life.[1] Writing was a great way to put the many cloudy hours at Palomar to good use. Gene was optimistically setting up the game plan, as he calls it, for the coming night's observing. Just in case. Philippe went outside to sample the brisk air.

Suddenly Carolyn sat up straight in her chair. She had spotted a strange object that at first didn't even appear to be real. "I went past something that could almost have been a floating edge-on galaxy," she recalls. "I knew that such a thing could not be possible, since only objects in our own solar system were near enough to float atop the starry background. So I moved the microscope stage back and took a second look. And there it was, a bar of light with coma and several tails!" The object stretched out so far that for a second she thought it could be some strange trail from an artificial satellite—except that it had appeared on both films. Because it appeared to float, she knew that the entire elongated thing was moving slowly through the sky.

Carolyn studied the images for a few more seconds. Then she looked up and said, "This looks like a squashed comet!"

The first of two discovery films of Comet Shoemaker–Levy 9. The bright ares at the film's perimeter are the result of damage done by someone who opened the box and exposed the films to light. Image by Gene and Carolyn Shoemaker and David Levy.

Surprised, Gene walked over, and Carolyn yielded the stereo-microscope to her husband. He studied the images carefully for a minute or so, and then he looked up toward me. In all the years I have known Gene Shoemaker, I have never seen such a look of pure bewilderment on his face.

Now it was my turn. I saw the images and quickly agreed they resembled a comet that someone had stepped on. Instead of a single coma and tail, there was a bar of coma, with a series of tails stretching to the north. But the weirdest part of this bizarre image was that on either end of the bar was a pencil-thin line.

By this time Philippe had returned, and we showed him the new find. Although we could not explain its strange appearance,

The discovery of Comet Shoemaker–Levy 9. These two images were taken by Gene and Carolyn Shoemaker and the author on March 23 (March 24 Universal Time) using the 18-inch Schmidt Camera at Palomar Mountain Observatory. The picture on the right was taken almost two hours before the one on the left.

we agreed it certainly did look like a comet of some sort. I dialed Brian Marsden's computer to send him a tentative discovery report. We were disconnected half way through, and so we re-sent an e-mail message:

Hi Brian,

We got cut off on our last message to you (the one we logged directly to your computer service) so we are resending with more details.

The strange comet is located as follows*:

1993 03 24.35503 12 26.7 (2000.0) −04 04 M = 14

---

*The comet's position numbers begin with the time: year, month, and date to five decimal places. The next figures are the comet's position in the sky. The 12 26.7 is the right ascension (celestial projection of longitude) and the −04 04 is the declination (latitude.) The 2000.0 means that the comet's position is referred to where the celestial coordinate system would be in the year 2000; because of precession it is always changing. Finally, the comet is 14th magnitude, probably about as bright as the planet Pluto.

The motion is west-northwest (not southeast as in the previous message) at about 7 arcminutes per day. The image is most unusual in that it appears as a dense, linear bar very close to 1 arcminute long, oriented roughly east-west. No central condensation [a thickening of the comet near its center] is observable in either of the two images. A fainter, wispy "tail" extends north of the bar and to the west. Either we have captured a most unusual eruption on the comet or we are looking at a dense tail edge-on.

Right now we are sitting in the middle of a cloud with no hope of observing tonight, and we had very poor observing last night. Observers are Eugene and Carolyn Shoemaker, David Levy, and Philippe Bendjoya.[2]

Brian Marsden replied quickly by e-mail. We could ask Rob McNaught, the indefatigable observer from Australia, to take a look. "Is there anyone else?" he asked.[3] There was. I knew that Jim Scotti was observing that very night from Kitt Peak's 36-inch Spacewatch camera. I suggested that we contact him. "Go ahead," Gene said.

Less than two hours after the discovery of Shoemaker–Levy 9, I recorded this note and sketch. There are two images of the comet since I was looking at both discovery films through the stereomicroscope and did not "merge" the two.

I worried that the storm we were in the middle of might have already reached Tucson. Maybe Eleanor Helin, whose team used the 18-inch just the week before, had captured the object on films they took. I checked the observatory's official log and found that Helin did have two films on one night and one film on a second night that could have included the object. We hoped to get confirmation from a telescope much larger than ours, a telescope capable of resolving the strangest cometary image our 18-inch had produced. We thought about Helin's film for a while. If Helin had not yet examined these films, our alerting her to the presence of the comet would mean that she could not then find it independently, and thus get credit for a discovery. We never thought that she and her team might have examined the films and yet failed to detect the comet, which seemed so bright to us.

It was now after 5 p.m., time to drive across the mountain the 5 miles to the house where we were staying so we could have our dinner and return to the dome before dark, again just in case the sky cleared. On the way home, Gene came up with a sudden insight: "What if this comet does not just appear to be near Jupiter in the sky? Suppose it really is near Jupiter? Maybe it recently passed within Jupiter's Roche limit and was disrupted." In other words, maybe Jupiter's gravitation tore the comet apart.

We all agreed that tidal disruption would explain the comet's barred appearance. We got to the house, where I made a sketch of the two discovery images. After dinner we returned to the dome. It was time to get our find confirmed. I called Jim, who said that he had received our message but added that he doubted that we had uncovered a real object. "In addition to its being close to Jupiter," he said, "it is also moving in exactly the same direction and at the same speed as Jupiter. I think you might have a strange reflection of Jupiter on your films."

I handed the phone to Gene, who discussed the possibility of a reflection—an artifact on the film—with Jim. Then Gene laid a straightedge along the bar and tried to see if it pointed straight back to Jupiter; this might indicate a reflection of some sort. It almost pointed to Jupiter, but not quite. Carolyn added that she

Carolyn Shoemaker, David Levy, and Gene Shoemaker at the 18-inch telescope dome at Palomar. Behind at right is the dome for the 200-inch telescope. Photo courtesy Ted Millis.

was sure the thing was real. "Let's have confidence in our discovery," I said. "Let's go measure it!"

Doing an accurate measurement of the position of a new comet relative to the known locations of 10 nearby stars is a significant task, for these positions are used to calculate the new object's orbit. In addition, the measuring engine at Palomar is a bear to use. "Madame Guillotine" I call it. We drove over to the 48-inch telescope, where we met Jean Mueller. An enthusiastic observer, Jean has been part of the 48-inch program since 1985. That program's goal is to photograph the entire northern sky to produce the second POSS, or Palomar Observatory Sky Survey. In the course of her work, Jean has discovered 10 comets, 72 distant exploding stars or supernovae, and several interesting asteroids. On March 25, 1993, she helped us measure the position of the new whatever-it-was.

Some two hours later, the measurement was done, and we returned to our dome. It was time to find out one way or another. Jim Scotti had said that even though he doubted that our object was real, he would try to look for it if the sky remained clear enough. We were worried, since the storm we were under was headed his way. I called the 36-inch telescope dome, and Jim answered. I could hardly recognize his voice.

"Are you okay?" I asked.

"Oh, yes. That sound you just heard was me trying to lift my jaw off the floor."

"Do we have a comet?"

"Boy, do you ever have a comet!" As Jim went on to describe the big-telescope image of the odd object we had found, I repeated his words to the others. The music of the fourth movement of Beethoven's first symphony playing on our stereo added to the drama. We decided to call this euphoria-inducing piece of music the Comet Symphony.

"There are at least five discrete comet nuclei side by side," Jim went on, "but comet material exists between them. I suspect that there are more nuclei that I'll see when the sky clears." As I repeated Jim's words, phrase by phrase, to the others, the excitement grew. Finally, Jim spoke about the wings. "On the east side

Using the 36-inch Spacewatch camera atop Kitt Peak, Jim Scotti took this remarkable image of Shoemaker–Levy 9 on March 30, 1993. Courtesy Jim Scotti and Spacewatch.

is a wing that's almost as long as the train of nuclei, and on the west side there's a much longer one!"

For a few seconds after I hung up the phone, we just stared at each other. "Well," Gene said, "what do you think of that!" Later, Gene said that he had to pull me down from the ceiling, but as I recall, all four of us were hovering pretty close to it. Philippe suggested that possibly this new object was not a comet but an asteroid that had somehow strayed too close to Jupiter. Although it definitely satisfied the definition of a comet and appeared, as the Greek definition of the word suggests, as a fuzzy or long-haired star, and although it did not have the look of a starlike asteroid, Philippe's off-the-cuff suggestion could have been correct. It is still too early to know where in the solar system this comet got started.

A moment like this doesn't happen very often in a comet hunter's career. We hadn't just found a comet; we'd found a real unicorn! Now Gene was ready to mention his tidal breakup theory to Brian, and we quickly penned a new message suggesting it. After Jim measured the comet's positions as shown in his images, he sent this message:

> Brian, here are my measurements of this remarkable object which the Shoemakers and David Levy have found. It is indeed a unique object, different from any cometary form I have yet witnessed. In general, it has the appearance of a string of nuclear fragments spread out along the orbit with tails extending from the entire nuclear train, as well as what looks like a sheet of debris spread out in the orbit plane in both directions. The southern boundary is very sharp, while the northern boundary spreads out away from the debris trails.
>
> Perhaps we can make arrangements to have you download a screen-dump later today. [Computerese for "We can send you a picture on computer."] It looks like the weather here is deteriorating with rain predicted for later in the day. I hope the system will clear out quickly, with only a night or two lost at most.

[Following this were Scotti's measured positions of the comet.]

Note 1: Nuclear region is a long narrow train about 47″ in length and about 11″ in width aligned along p.a. 260–80 degrees. [Jim's measurement of comet's size and orientation.] At least 5 discernible condensations are visible within the train, with the brightest being about 14″ from the SW end of the trail. Dust trails extend 4.20′ in p.a. 74 degrees and 6.89′ in p.a. 260 degrees, roughly aligned with the ends of the trail, measured from the midpoint of the train. A tail extending more than 1 minute from the nuclear train with the brightest component extending from the brightest condensation in the train to 1.34′ in p.a. 286 degrees. The midpoint of the train was used for the astrometric measures.

Note 2: Observations through cirrus.

Time for bed!

Jim, Mar. 26, 14:17 UT.[4]

The time 14:17 was universal, or Greenwich, England, time; that means he sent his message at 7:17 a.m. Mountain Standard Time. He had been up all night despite the cirrus clouds, and he stayed up still longer to send his note. We stayed up most of the night too, just in case the sky cleared, but we were too excited to leave early in any case. But once we knew the likelihood of further observing was completely hopeless, we went home.

THE AFTERMATH

The morning of March 26 dawned foggy and cold. I was up pretty early, but when I opened my door I saw Carolyn already about. She turned to me, grinned broadly, and gave me a big hug. As we talked about our find, she admitted that it was the most exciting discovery in all her years of observing.

Even before Brian Marsden's office issued the discovery circular (which is reproduced on the next page), news about the strange comet was buzzing about. At the Lunar and Planetary

Circular No. 5725
Central Bureau for Astronomical Telegrams
INTERNATIONAL ASTRONOMICAL UNION
Postal Address: Central Bureau for Astronomical Telegrams
Smithsonian Astrophysical Observatory, Cambridge, MA 02138, U.S.A.
Telephone 617-495-7244/7440/7444 (for emergency use only)
TWX 710-320-6842 ASTROGRAM CAM   EASYLINK 62794505
MARSDEN@CFA or GREEN@CFA (.SPAN, .BITNET or .HARVARD.EDU)

COMET SHOEMAKER-LEVY (1993e)

Cometary images have been discovered by C. S. Shoemaker, E. M. Shoemaker and D. H. Levy on films obtained with the 0.46-m Schmidt telescope at Palomar. The appearance was most unusual in that the comet appeared as a dense, linear bar $\sim 1'$ long and oriented roughly east-west; no central condensation was observable, but a fainter, wispy 'tail' extended north of the bar and to the west. The object was confirmed two nights later in Spacewatch CCD scans by J. V. Scotti, who described the nuclear region as a long, narrow train $\sim 47''$ in length and $\sim 11''$ in width, aligned along p.a. $80°$-$260°$. At least five discernible condensations were visible within the train, the brightest being $\sim 14''$ from the southwestern end. Dust trails extended 4.20 in p.a. $74°$ and 6.89 in p.a. $260°$, roughly aligned with the ends of the train and measured from the midpoint of the train. Tails extended $> 1'$ from the nuclear train, the brightest component extending from the brightest condensation to 1.34 in p.a. $286°$. The measurements below refer to the midpoint of the bar or train.

1993	UT	α_{2000}	δ_{2000}	m_1	Observer
Mar.	24.35503	$12^h 26^m 39.27$	$-4°03'32.9$	14	Shoemaker
	24.43072	12 26 37.21	-4 03 23.0		"
	26.29531	12 25 42.24	-3 57 55.7	13.9	Scotti
	26.30479	12 25 42.09	-3 57 53.7	16.7	"
	26.31448	12 25 41.63	-3 57 53.7		"
	26.41291	12 25 38.70	-3 57 34.8		"

C. S. Shoemaker, E. M. Shoemaker, D. H. Levy and P. Bendjoya (Palomar). Measurers D. H. Levy, J. Mueller, P. Bendjoya and E. M. Shoemaker. J. V. Scotti (Kitt Peak). Last observation made through cirrus.

The comet is located $\sim 4°$ from Jupiter, and the motion suggests that it may be near Jupiter's distance.

SUPERNOVA 1993E IN KUG 0940+495

D. D. Balam and G. C. L. Aikman report a measurement of $V = 20.3 \pm 0.1$ and $B - V = +0.51$ on Feb. 26.28 UT, using the 1.85-m reflector ($+$ CCD) at the Dominion Astrophysical Observatory.

1993 March 26   *Brian G. Marsden*

Circular *5725 of the International Astronomical Union announces the discovery of the comet that would eventually be known as Shoemaker–Levy 9.*

Laboratory in Tucson, people stood around computer screens displaying Jim's image of the comet. Within a few hours, reports of other observations began coming in from around the world. Brian Marsden issued a preliminary orbit computed under the assumption that the comet had passed near Jupiter eight months earlier, on July 28, 1992—a prescient notion that turned out to be uncannily close to correct.

Within a few days other observations were announced. Observers who had taken films earlier but who had not yet examined them, or who had examined them and missed identifying the new comet, sent their reports also. One observation was from as early as March 15. The advantage of these early data was that Marsden was able to calculate our strange comet's orbit within two days of the discovery.

Because the number of precise positions reported was low and covered only a brief time, this early orbit calculation was far from accurate. Another thing puzzled us: Although there was a strong possibility that Jupiter had disrupted the comet, the giant planet's guilt was not established. Comets can fall apart for lots of reasons, Scotti said in the first of many press interviews he would have about the comet. "You can sneeze on a comet and it will fall apart."[5]

THE FIRST WEEKEND

With a cloudy weekend upon us, we had a chance to relax and reflect on our comet. We heard that Richard McCrosky had observed it from Harvard's 61-inch reflector at Oak Ridge, Massachusetts, and that he thought the comet looked like a meteor frozen against the sky. Then Brian told us that Jane Luu and David Jewitt had used the University of Hawaii's 88-inch reflector and a large CCD to take a magnificent image of the comet. "We resolved 17 separate subnuclei strung out like pearls on a string," they wrote, "along a position angle 77 degrees [where 90 degrees would be toward the east]. The distance between first and last nuclei in the string is 50 arcsec. [Fifty seconds of arc is almost one-thirtieth the diameter of the Moon in the sky.] Your IAU circulars understate the impressive nature of this comet![6]

Of all the early labels for this comet, Jewitt's "string of pearls" was the most elegant. Besides calling it a "squashed comet," Carolyn also thought the comet looked like "lights on a spaceship"—a more attractive image than Brian Marsden's choice of metaphor: "a mouth with teeth."

A CONTROVERSIAL DISCOVERY

Whether discovery involves a concept, or an object, it usually very much involves the discoverer's gut.

When Henry Holt and I discovered the first asteroid locked in an orbit that kept it a certain distance from Mars—the first *Mars Trojan* asteroid—we proposed the name "Eureka" for it. The name comes from the apocryphal story of Archimedes. Challenged to determine which of two crowns was made of real silver, he figured out his now famous principle of objects displacing different amounts of water while he was taking a bath. He was so excited that he leaped out of the tub, presumably still unclad, and ran out of the house shouting "Eureka!" (I found it!)

On February 18, 1930, Clyde Tombaugh discovered Pluto. He knew instantly that he had found something in the solar system far beyond the planet Neptune. There was no doubt in his mind. He spent the next 45 minutes running tests on his find to make sure that the object's motion confirmed its great distance. "You have to have the imagination to recognize a discovery when you make one," Tombaugh said. "I would suggest that above everything else, in observing you have to be very alert to everything. You have to be able to recognize a discovery as such."[7]

On April 1 and April 3, Marsden issued circulars about our new find. They contained "prediscovery" observations, one by Endate on March 15, and the other by Otomo two nights later.

For amateur and professional astronomers deep into comet seeking, missing a prize is a keen disappointment. Comet hunting is a game that is part science, part art, and part competitive sport. But it must be the world's slowest sport. People can spend years searching for the elusive fuzzy prey, only to have it rise behind a tree or a cloud on the critical night. All successful finders have had this painful experience.

Observing from La Silla, Chile, at the European Southern Observatory, Mats Lindgren was searching for faint comets in orbit about Jupiter when he picked up Shoemaker–Levy at the end of March and promptly dismissed it as an artifact, albeit a very bright artifact, on his photographic plate.[8]

From films her team had taken a week before our discovery, Eleanor Helin reported that the comet was 12th magnitude— dazzlingly bright for the 18-inch. Since her films were neither fogged nor obscured by clouds, we were not surprised that the comet would look so bright. But then Marsden added a puzzling statement: "The comet was noted at the time." How? If Helin had indeed identified an object that looked like a comet, and it had been observed on two nights, why hadn't they reported it to Marsden? "Helin was working frantically on a proposal," Marsden explained to us. "She didn't check the second night's image until after I announced your discovery."

Meanwhile, back for another observing session at Palomar, Helin told a colleague that she had simply taken the object to be either a galaxy or an artifact. Later, she clarified this at a meeting held in Sicily in June. Her pair of films, she explained, were taken only 36 minutes apart, not enough time for her to see the comet "floating" above the starry background as a stereomicroscope will make a moving object do. One of her images, published in *Southern Sky*, showed that the telescope had not been guided well during the exposure and that all the images—stars and comet— were trailed. Thus, Helin would not have noticed the comet appearing to float above the background of stars.[9]

Finally, on March 26, Orlando Naranjo began a series of three exposures from the Llano del Hato Observatory in Merida State in Venezuela. On March 30 he reported three nights of observations of Shoemaker–Levy 9.[10] "When I saw the diffuse images in my

On March 28, 1993, Wieslaw Wisniewski (Univeristy of Arizona) used the 90-inch telescope at Kitt Peak to record the newly discovered comet. Courtesy W. Wisniewski and J. V. Scotti, University of Arizona.

plates," Naranjo writes, "I had not heard about Shoemaker–Levy." He did notice that the object was moving at about the same rate and direction as Jupiter.[11]

Each such story represents extreme disappointment for these dedicated teams and individuals, so I don't mention them casually. I love comet hunting more than I can ever say, and I feel that the sport should be a happy one, all the time, for everyone with the persistence to search. Eleanor Helin's program at Palomar has been remarkable, as the many comets that orbit the Sun bearing her name testify.

THE 18-INCH SCHMIDT

Although the new comet that Carolyn, Gene, and I discovered that fateful night is probably the most unusual discovery ever made through Palomar's workhorse 18-inch telescope, it is only one of a very long series that began soon after the telescope first saw light from the stars in 1936. It was Russell Porter's first effort at Palomar. An Arctic explorer, artist, and amateur astronomer, Porter in 1925 launched a telescope makers' conference near Springfield, Vermont. Porter called the site of the conference "Stellar Fane" for Shrine to the Stars, and the name was later compressed to Stellafane.[12] In November 1925 a *Scientific American* article extolled amateur telescope making as a way of beating the high prices of commercial refractors.[13] But more than that, the grinding of mirrors was touted as an adventure that begins in a basement with glass and grit and ends in the stars.

By 1930, Porter's reputation had spread across the country and attracted the attention of George Ellery Hale, who was completing the design for a 200-inch reflector telescope that would be the largest on Earth. Hale persuaded Porter to abandon the green hills of Vermont and head for California. Once there, Porter designed and built several beautiful 4- and 12-inch-diameter telescopes meant to find the best location for the observatory. The 18-inch was apparently his first telescope project to be mounted permanently at Palomar. Glad to return to "pushing glass," Porter completed the optical figuring of the 18-inch.[14] The final design of the 200-inch reflector incorporated a mount that looked like a giant horseshoe, similar to the small horseshoe-shaped stand of an elegant garden telescope that Porter had previously designed.

The 18-inch telescope was the brainchild of Fritz Zwicky, a bear of a man who could be heard two blocks away when he was chatting in a hallway.[15] He used the 18-inch to search for the exploding stars, or supernovae, in distant galaxies. With this telescope and the larger 48-inch Schmidt, Zwicky found more than 150 supernovae. The telescope became famous for comets under Gene Shoemaker, whose monthly searches with Carolyn have yielded 32 comets, and Eleanor Helin, whose program has discovered 15 comets. These teams have found countless asteroids using the 18-inch, including a few whose orbits cross that of the Earth and so will pose a danger to our planet at some far-off date in the future.

The 18-inch was also used to discover an object that, in a writer's imagination, swallowed the Earth and held it hostage for several months! The pages of *The Black Cloud*, a 1950s novel by the British astronomer Fred Hoyle, recount a tale of astronomers spotting a large dark cloud through Palomar's 18-inch Schmidt. They identified the object as a Bok Globule, a dark cloud of gas and particles of dust that the Milky Way specialist Bart Bok thought were the birthplaces of new stars. But this one, larger than the others, was headed straight for Earth. Two months later it swallowed us, leaving a pitch black sky. After scientists finally understood that the globule was, incredibly, an intelligent creature, they were able to persuade it to leave Earth. The novel was quite popular in its day; it educated the public about the new types of objects that astronomers were finding and the telescopes used for finding them.

But that was 1955. Almost four decades later, this same telescope recorded an object that was just about as strange. Although it was not the first time a comet had been seen to break up into pieces, Shoemaker–Levy was by far the most dramatic example of a disrupted comet ever found. As our observing run that March finally came to an end, the sky cleared slightly the last hour before dawn. We loaded a film into that venerable 18-inch Schmidt telescope and took a photograph of our comet, which had now moved considerably from its discovery position.

We had no idea that cold March night that the comet's dramatic appearance was only a preamble.

A Scorpion's Embrace

In the months after the discovery, a series of surprises kept up our excitement level. The first, an observational one, was that the comet was disrupted into so many pieces. The other bombshells were mathematical, revealed only after our understanding of the comet's orbit deepened. Surprise No. 2: That the comet had passed very close to Jupiter, allowing its gravitational pull to tear it apart. Surprise No. 3: That the comet had abandoned its direct solar orbit and become a satellite of Jupiter. And, finally, Surprise No. 4: That the comet's orbit was its last, and that it would, in fact, collide with Jupiter.

A DISRUPTED COMET AND CHAINS OF CRATERS

In the days after our discovery, astronomers speculated on how the comet could have taken on its strange appearance. It seemed as though some disruption by Jupiter was the most likely possibility. But an important clue came from the Jovian moons,

Callisto and Ganymede, and, curiously enough, our own Moon. That story began in 1979, when the Voyager spacecraft revealed the existence of several chains of craters on two of Jupiter's moons. On the ancient and scarred surface of Jupiter's moon Callisto, Voyager photographed 13 long chains, or catenae, of craters. The chains vary in length from 200 to 650 kilometers. One straddles the rim of Valhalla, a large-impact basin. Gipul Catena, one such chain, consists of about 20 equal-sized impact sites that total 650 kilometers in length. At the time of Voyager's discovery, planetary geologists Quinn Passey and Gene Shoemaker speculated that these craters were formed from the debris sent off by a larger impact somewhere else on Callisto. Ganymede, with a somewhat younger surface, has three such chains. The much younger surfaces of Europa and Io have no craters of any sort. There is even a small chain of craters on our own Moon. Splattered across the large crater called Davy, this Davy Catena consists of some 20 small-impact sites neatly lined up in a straight line.

A few days after the comet's discovery, Jay Mellosh, a planetary astronomer at the University of Arizona's Lunar and Planetary Lab, was casually reading his morning paper when he saw Wieslaw Wisniewski's beautiful photograph of comet fragments lined up together. Mellosh recalled this unsolved mystery from the Voyager spacecraft of 15 years earlier: Mellosh thought of these chains as he looked at the newspaper picture of the disrupted comet. Could these sites be records of ancient versions of Shoemaker–Levy 9 that broke up as they rounded Jupiter and then plowed into a Jovian moon?[1] The statistics seemed right: Over the long history of the solar system, objects would fall apart near Jupiter often enough that some tiny percentage should hit the moons. Finally, all but three of these chains are on the side of Callisto facing Jupiter, and those three are not far from the Jupiter-facing side. In the case of the Davy Catena, a comet or a loosely held together asteroid probably grazed the Earth. Passing just outside the atmosphere, the object fell apart, its pieces racing away to crash into the Moon perhaps a day later. For such a cosmic coincidence to happen just once, the Earth would have to have

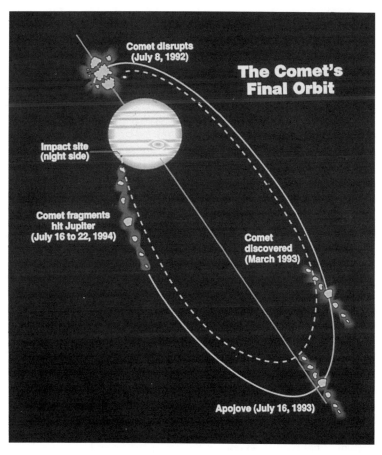

Comet Shoemaker–Levy 9's final orbit. The comet broke apart on July 8, 1992, and struck Jupiter two years later. The orbit, comet, and planet are not shown to scale, and the final cigar-shaped loop was more elongated than shown here. Courtesy D. Yeomans, P. Chodas, and Sky and Telescope.

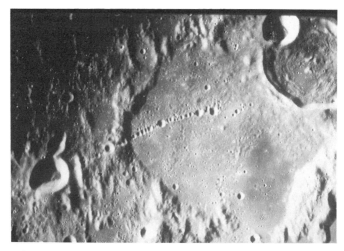

More than a billion years ago, a comet brushed by Earth and broke apart from tidal stress. Probably within a day or two, the pieces of the shattered object collided into an ancient crater on the Moon, a crater we now call Davy. Until the University of Arizona's Jay Mellosh made the connection with Shoemaker–Levy 9, astronomers had no idea why chains of craters like this one appeared on our Moon and on Jupiter's moons Ganymede and Callisto.

been grazed by hundreds of kamikaze comets, all in Earth's distant past. Most of these left our neighborhood, never to return or to come back millennia later and collide. Some might have done a final loop around the Earth before colliding with our planet. One made it to the Moon.

Figuring out how the comets became disrupted would help explain what happened both to the crater chains and to Shoemaker–Levy 9. Mellosh posited a model where the comet is originally fitted with material held together end to end like a stack of coins. As the comet passes Jupiter, the coins separate into a line of material.

If a handful of disrupted comets smashed into Callisto, Mellosh assumed that some of the rest would either leave Jupiter's sphere of influence and continue as separate comets in disparate paths or that some could return to crash into Jupiter.

With the exception of the discovery itself, all these revelations came through celestial mechanics. Using this beautiful mathematical field of study to examine a comet begins with simply photographing it on many nights. Then comes the task of astrometry, which involves measuring the comet's positions with respect to the known locations of background stars. From at least three such positions, a celestial mechanician like Brian Marsden, Donald Yeomans at Caltech's Jet Propulsion Laboratory, or Japan's Syuichi Nakano can calculate a preliminary orbit. The more positions that are available, especially spread out over a long period of time, the more precise the orbit calculation.

While this procedure works for typical comets, Shoemaker–Levy was anything but typical. It appeared to be close to Jupiter, but how close? An intelligent computation, Brian notes, had to try to include the gravitational attractions not just by the Sun but also by Jupiter, which was difficult, since the comet's distance from Jupiter was unknown.

TIDAL DISRUPTION

From the day the comet was found, we suspected that a previous encounter with Jupiter had brought about the tidal disruption that made it fall apart. Anyone who lives near an ocean sees how the gravitational pull of the Moon, and to a smaller extent the Sun, causes a gentle surge of water from a low level to a higher level and back again as Earth rotates. Because of unusual hydrologic effects, the tides in Nova Scotia's Minas Basin are anything but gentle; the tides there can rise or fall 50 feet in just a few hours. (See Chapter 15.)

Even a 50-foot tide is tiny compared to what would happen if a small object came to within a certain distance—the Roche limit—of a planet. This mathematical limit, defined by Roche, applies to a body that has nothing but its own gravity to hold it together, like a sphere of water. Although hundreds of artificial satellites orbit within the Earth's Roche limit without falling apart, these "moons" are made with metal and bolts. We live well within the Roche limit, on the Earth's surface, and we do not fall apart. But water spilled out of a glass will not retain its shape; it will flow. Imagine, though, a glass spilled in space far away from Earth and far from any planet's Roche limit. That water would stay together as a sphere.

The easiest way to explain the presence of a completely disrupted comet is to say that it must have passed within Jupiter's Roche limit. But since the early orbits did not show the comet passing that close to the giant planet, other ideas were advanced. One suggested that the comet had rotated so rapidly that its own angular momentum tore it to pieces. There are examples of comets splitting apart for no apparent reason. In 1988 Comet Shoemaker–Holt chased Comet Levy, my fourth comet find, around the Sun, losing the race by 76 days. Brian Marsden noted that these two

On March 27, 1993, Jane Luu and David Jewitt captured Comet Shoemaker–Levy 9 using the 88-inch telescope atop Mauna Kea, Hawaii. Courtesy Jane Luu and David Jewitt.

comets were one, and some 12,000 years ago they split apart. So gentle was the split, however, that the comets might have orbited each other for a while before finally moving off on their own.[2]

Other comets fracture far more dramatically. More than a century ago, in 1888, the German astronomer Heinrich Kreutz suggested that a series of bright comets that did hairpin turns about the Sun actually shared a common origin. His notion might have been inspired by a large comet that did a hairpin turn round the Sun in 1882, then breaking into several pieces. Another great comet, Ikeya–Seki, came back as the Great Comet of 1965 and split in two. Two years later, Marsden established that these two comets had visited before as a single comet, and that this comet was at perihelion sometime during the first half of the twelfth century. He suggests that the sun-grazing comet of 1106 might have been this progenitor comet.

How far back can we trace this group? A bright comet that appeared in 373 B.C. was accused of flooding the towns of Helice and Buris. Seneca described this comet in his *De Cometis*, the only surviving record of a story told by the fourth century B.C. Greek astronomer Ephorus. Seneca couldn't believe his ancient colleague's account of the comet's "splitting into two stars."[3] "It requires no great effort to strip Ephorus of his authority," Seneca huffed, since apparently no one else reported the comet's division. But modern celestial mechanics has shown that Ephorus was probably right and that his observation was a valuable one.

If comets can split apart near the Sun, they can also crumble, as Periodic Comet Brooks 2 did in 1886, after a close approach to Jupiter. (See Chapter 1.)

CHRISTIAAN HUYGENS AND THE RINGS OF SATURN

In 1610, when Galileo turned his telescope toward Saturn, he expected to see moons similar to those he had discovered near Jupiter. He didn't. To his shock the planet seemed to be surrounded by two large objects that, after a few years passed by, apparently faded away. Galileo never figured out this mystery of Saturn. That was left to his follower Christiaan Huygens, a brilliant Dutch scientist. Huygens was prodigious in his accomplishments. He was the first to use a pendulum to run a clock and developed the theory that light travels in waves. Eyeing Saturn through his telescope in 1655, he saw something so incredible he decided to publish it as an anagram. When deconvolved, the Latin anagram read, "*Annulo cingitur, tenui, plano, nusquam coherente, ad eclipticam inclinato.*" In this roundabout way the world learned that Saturn was "surrounded by a thin, flat ring, nowhere touching, inclined to the ecliptic."

More than three centuries after their discovery, estimates still differ on how old the rings are. Although one investigator has them as young as 100 million years old, most scientists think that the rings formed very early in the history of the solar system. According to studies of the Voyager data of Saturn, the rings could have been formed by the collisional disruption of three satellites.[4]

Because, as Huygens so presciently observed, Saturn's rings are inclined to the ecliptic, we see them differently each observing season as Saturn travels round the Sun. In 1966, 1980, and 1995, for example, the rings presented themselves edgewise so that we could not see them at all.

When the Voyager spacecraft visited Saturn in 1980, they found that each of the major rings was subdivided into hundreds of thin ringlets looking somewhat like grooves in a record. Small moons actually shepherd particles into these thin but discrete rings. In 1989, the planet and rings passed in front of 28 Sagittarii. It was a thrill to watch the star slowly blink on and off as the rings paraded past.

Although Saturn's rings are the most dramatic example of what happens after objects break up within a planet's Roche lobe, they are not the only ones. In 1978 observers timing an occultation of a star by Uranus detected a ring system around the distant green giant planet. While visiting Jupiter in 1979, Voyager 2 found a faint ring surrounding the gas giant. Ten years later, the same spacecraft found an assortment of rings around Neptune.

JUPITER'S 21 NEW SATELLITES

Little more than a week after the discovery, Brian Marsden published a new IAU circular about the comet. "Attempts at orbit determination from the Mar. 15–Apr. 1 arc," he noted, "suggest that a parabolic solution is no longer viable." Brian always believed that this comet was in some form of periodic orbit; his initial parabolic orbit was published as little more than a formality. This meant that the comet was not on a long one-time-only look at the Sun and the inner planets. He went on to suggest that the comet had a very close approach to Jupiter sometime during 1992. His conclusion: "The object is at least temporarily in orbit about Jupiter."[5]

Our comet in orbit around Jupiter? We were surprised and pleased with Brian's calculation. With most comet finds, there is an initial surge of adrenalin at the discovery, followed by a second surge when the orbit is announced. But with this comet's list of surprises, our adrenalin high was ongoing.

With each passing week, observations flooded in from observatories around the world. Most of the observations were made during the two weeks of each month when the sky was not floodlit by moonlight. By April 10, Brian was still not convinced about the comet's orbit. Although he strongly suspected it, he cautioned that "we really don't know that the object is currently a satellite," adding that computations of an orbit about Jupiter are very difficult since we still do not know how far the comet is from that planet.

At the end of April I traveled to Cambridge, Massachusetts, to give the first of what became a long series of lectures about the comet. Whenever I went to Cambridge, I visited Brian and his associate, Dan Green. On this Monday, April 19, 1993, Dan and I had joined Steve O'Meara from the nearby offices of *Sky and Telescope* to observe an enchanting celestial event: the Moon passing in front of Venus. Even though the event took place in the middle of the day, we were able to see it. As Steve peered through the rooftop telescope, he suddenly shouted, "That's it!" We all saw the incredible sight of a crescent Venus emerging from behind our crescent Moon. Afterward I visited Brian, who was busily finishing off the comet's latest orbital calculations. He planned to publish these in the next issue of the *Minor Planet Circulars*.[6] By now his previous uncertainty had vanished. "The comet was definitely in orbit about Jupiter," he said, which meant that the gas giant was also carrying the comet along in its orbit about the Sun. It was time, Brian decided, to give the comet a new designation. Since the comet was a periodic comet—a comet defined as orbiting the Sun in a period 200 years or shorter—it would be renamed in the tradition of periodic comets, with a number that reflects periodic comets bearing the same name. It was true that the comet was orbiting Jupiter, but since Jupiter orbits the Sun, the comet could be said to orbit the Sun also, just as our Moon, though orbiting the Earth, also orbits the Sun. So the latest Shoemaker–Levy was now Periodic Comet Shoemaker–Levy 9, abbreviated S-L 9. There are five other Shoemaker–Levy comets in far larger orbits; these have no numbers.

THE BIGGEST SURPRISE: IMPACT

Near the end of April, a number of scientists went to Erice. During that meeting Brian Marsden and Japan's Syuichi Nakano were trading e-mailed combinations of different orbits that would fit the observed positions of the comet. "While I was there," Brian recalls, "among the several orbits Nakano and I were computing, he got one that took the comet in really close to Jupiter sometime during a two-week interval in June–July 1992. 'That's it!' I said."[7]

By the end of the second month after discovery, there were enough observations to clarify two things about the comet's orbit about Jupiter. The first was that on July 7, 1992, the comet had passed to within 50,000 kilometers of the planet, astoundingly close to the gas giant and well within Roche's limit. It was now the middle of May 1993, and the Shoemakers and I were about to leave for Palomar. I checked my electronic mail to see if any interesting visitor, like a new comet or asteroid, had been discovered recently.

We would want to know about such objects so that we could observe them and measure their positions. Without observations such as these, new objects could be lost.

While checking my e-mail, I sent Brian a message, adding that we looked forward to a good run. Brian's response included the surprising information that the latest orbits for S–L 9 showed an increasing likelihood that the comets would collide with Jupiter. For a moment I thought about how rare and exciting a collision would be, but since the thrill of getting back to Palomar was more immediate, I quickly forgot Brian's words.

MAY 22, 1993

Even at a remote mountaintop like Palomar, Saturday afternoons are quieter than the rest of the week. When we arrived at the dome that day, I looked forward to a relaxing afternoon of writing and preparing for what would surely be a clear night. Since the night before had been cloudless, keeping us up observing until dawn, the Shoemakers and I awoke well past noon. After our breakfast, we made our way back to the observatory and the 18-inch telescope dome.

The first thing I did that afternoon was switch on my laptop and modem and plug into the outside world to see if anything new had happened—not new on Earth, but new in the sky. Just as in Tucson, I wanted to know if any new objects had come along that we needed to know about. Meanwhile, Carolyn used her stereomicroscope to scan the previous night's photographic bounty, and Gene was in the darkroom preparing film for the coming night's observing.

As soon as I logged on, I knew that something was up. "There are two new circulars about S–L 9!" I called out. I remember when someone asked Brian how he would announce that something was on a collision course with Earth. "I would just publish the orbit and ephemerides," he said. "Readers could then see that Delta [distance of the object from the center of the Earth] was

Circular No. 5800

Central Bureau for Astronomical Telegrams
INTERNATIONAL ASTRONOMICAL UNION
Postal Address: Central Bureau for Astronomical Telegrams
Smithsonian Astrophysical Observatory, Cambridge, MA 02138, U.S.A.
Telephone 617-495-7244/7440/7444 (for emergency use only)
TWX 710-320-6842 ASTROGRAM CAM   EASYLINK 62794505
MARSDEN@CFA or GREEN@CFA (.SPAN, .BITNET or .HARVARD.EDU)

PERIODIC COMET SHOEMAKER-LEVY 9 (1993e)

Almost 200 precise positions of this comet have now been reported, about a quarter of them during the past month, notably from CCD images by S. Nakano and by T. Kobayashi in Japan and by E. Meyer, E. Obermair and H. Raab in Austria. These observations are mainly of the "center" of the nuclear train, and this point continues to be the most relevant for orbit computations. Orbit solutions from positions of the brighter individual nuclei will be useful later on, but probably not until the best data can be collected together after the current opposition period. At the end of April, computations by both Nakano and the undersigned were beginning to indicate that the presumed encounter with Jupiter (cf. *IAUC* 5726, 5744) occurred during the first half of July 1992, and that there will be another close encounter with Jupiter around the end of July 1994. Computations from the May data confirm this conclusion, and the following result was derived by Nakano from 104 observations extending to May 18:

$$\text{Epoch} = 1993 \text{ June } 22.0 \text{ TT}$$

$$
\begin{aligned}
T &= 1998 \text{ Apr. } 5.7514 \text{ TT} & \omega &= 22\overset{\circ}{.}9373 \\
e &= 0.065832 & \Omega &= 321.5182 \\
q &= 4.822184 \text{ AU} & i &= 1.3498
\end{aligned} \Bigg\} 2000.0
$$

$$a = 5.162007 \text{ AU} \quad n° = 0.0840381 \quad P = 11.728 \text{ years}$$

This particular computation indicates that the comet's minimum distance Δ_J from the center of Jupiter was 0.0008 AU (i.e., within the Roche limit) on 1992 July 8.8 UT and that Δ_J will be only 0.0003 AU (Jupiter's radius being 0.0005 AU) on 1994 July 25.4.

As noted on *IAUC* 5726, the positions of the ends of the nuclear train can be satisfied by varying the place in orbit at the time of the 1992 encounter and considering the subsequent differential perturbations. Using the above orbital elements, the undersigned notes that the train as reported on *IAUC* 5730 corresponds to a variation of ±1.2 *seconds*. Separation can be regarded as an impulse along the orbit at encounter, although the velocity of separation (or the variation along the orbit) depends strongly on the actual value of Δ_J. At the large heliocentric distances involved any differential nongravitational acceleration must be very small, as Z. Sekanina, Jet Propulsion Laboratory, has also noted. Extrapolation to shortly before the 1994 encounter indicates that the train will then be ∼ 20′ long and oriented in p.a. 61°-241°, whereas during the days before encounter the center of the train will be approaching Jupiter from p.a. ∼ 238°.

1993 May 22   *Brian G. Marsden*

Circular *5800 announces that Comet Shoemaker–Levy 9 will pass "only 0.0003" astronomical units from the center of Jupiter, "Jupiter's radius being 0.0005 AU"—in plain English, there will be a collision. An astronomical unit is the distance between Earth and Sun.*

decreasing to the Earth's radius." On Circular 5800, he had the opportunity to do just that, but thankfully for a planet other than Earth.

Brian Marsden's circulars never read like newspapers: "Comet to Strike Jupiter!" isn't official Central Telegram Bureau style. Instead the item, which was simply headed "Periodic Comet Shoemaker–Levy 9 (1993e)," described the comet's final orbit:

"At the end of April," Brian explained, "computations by both Nakano and the undersigned were beginning to indicate that the presumed encounter with Jupiter (cf. IAUC 5726, 5744) occurred during the first half of July 1992, and that there will be another close encounter with Jupiter around the end of July 1994."

Then Brian explained the details of the new orbit Nakano had calculated: Delta_J means the distance from Jupiter, expressed in AU (where 1 AU, or astronomical unit, represents the mean distance between the Earth and the Sun). "This particular computation indicates that the comet's minimum distance Delta_J from the center of Jupiter was 0.0008 AU (i.e., within the Roche limit) on 1992 July 8.8 UT and"—here is the sentence he had long wished to write—"that Delta_J will be only 0.0003 AU (Jupiter's radius being 0.0005 AU) on 1994 July 25.4."[8]

It was easy enough to translate this message. Although the comet fell apart in 1992, its pieces survived the graze with Jupiter, but only to buy a little time. The ancient comet would have one orbit left, a last chance to swing away from Jupiter, look back, and return again to crash into the planet.

"Carolyn," I said, "our comet is gonna hit Jupiter."

At first Carolyn didn't seem at all pleased that our comet was about to pull a kamikaze act. But at this moment the many phases of Gene Shoemaker's long career were about to come together. His Ph.D. thesis demonstrated that Meteor Crater, the big landmark east of Flagstaff, Arizona, was the result of an impact of an asteroid some 50 meters wide. In the 1960s he studied the impact craters on the Moon as seen through the telescope and the "eyes" of the *Ranger*, *Surveyor*, and *Apollo*. These craters were caused by prehistoric hits by comets and asteroids. For several months of

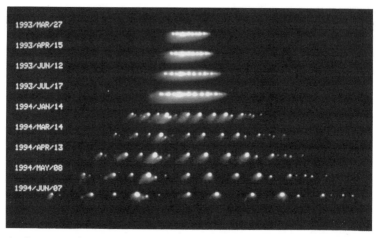

1993/MAR/27
1993/APR/15
1993/JUN/12
1993/JUL/17
1994/JAN/14
1994/MAR/14
1994/APR/13
1994/MAY/08
1994/JUN/07

An observer standing on one of the comet fragments might see the others moving apart at about a runner's pace—about 3 meters per second. This sequence shows the fragments continuing to distance themselves with the passage of time.

each of the past 10 years, Gene and Carolyn have studied impact craters in the Australian outback. At Palomar, they searched for the asteroids and comets that forge the craters when they crash into other worlds.

Now from outside the darkroom, Gene was hearing me say that our comet was going to plummet into Jupiter.

I was still trying to read the circular aloud. "What! What'd he say?" Gene yelled from inside the darkroom. I could hear lids slam shut and drawers close as he quickly tried to put film away. Then he thrust the door open, ran outside, pushed me out of the way, and read the computer screen intently. "I never thought I'd live to see this!" he whispered. "We're going to see an impact!"

Gene realized that Jupiter's enormous gravity had a better chance than Earth's gravity of bringing in a comet. In fact, he believed that at least one comet larger than a kilometer in diameter was probably in a temporary orbit about Jupiter at any given time.

But at some 500 million miles from us, such a comet would normally be a frozen ball of mud far too faint to see. We usually learn about comets orbiting Jupiter only after they have left its sphere of influence and come close enough to the inner solar system, Earth's neighborhood, for us to see them.

If Brian's bombshell wasn't enough for one day, that very night we turned the telescope toward the far southern constellation of Centaurus, long ago named for the rowdy creatures with bodies of horses and heads of people. The films we took there contained yet another new comet, which would be called Shoemaker–Levy and designated 1993h. This one was on a very wide parabola of an orbit that brought it in from a great distance, perhaps as far as a light year from the Sun in a place called the Oort cloud of comets. The comet would soar once round the Sun and back away tail first, not to return for a very long time.

Such would not be the fate of Shoemaker–Levy's more troubled sister. Caught in an ever narrowing orbit about Jupiter, it traveled under a death sentence. There was an excellent chance that in 14 months, it would end its fractured life in a crash against the mighty planetary giant.

About Comets

It is amazing that the solar system could create an object so fragile, and that it would stay together for so long.
—Donald Yeomans, Jet Propulsion Lab[1]

Periodic Comet Shoemaker–Levy 9 was about as far from being a normal comet as you could get. Its resemblance to a string of pearls and its orbit about Jupiter made it unlike anything we have seen in the comet world. But to appreciate a very unusual comet, it helps to know a little about what garden variety comets look like and how they behave.

KEPLER'S COMET AND THE ONSLAUGHT OF 1618

A series of at least three bright comets rounded the Sun in 1618. The first was independently found by the legendary astronomer Johannes Kepler from Linz, Austria. An energetic observer, he discovered the comet in the morning sky before sunrise. Although this comet is not officially known as Kepler, his independent sighting and the fact that we do not know the names of the Hungarians who had found it two days earlier suggest that we could informally name it Comet Kepler after him.

A second resplendent comet came along just two months later, again in the morning sky, and a third after that. "Actually," noted Brian Marsden, "it was at one time thought that there were five comets, but it would seem that even four is an overestimate."[2] It is possible that two of the others were really the other comets seen in different parts of the sky after they rounded the Sun. The important thing about this cometic stampede is its timing just eight years after Galileo first turned his telescope to Jupiter and discovered its four moons. It sparked an intense debate on the nature of comets. Were they part of our atmosphere, or did they orbit the Sun like the Earth did? The eminent Galileo himself took part, coming out of a self-imposed retirement to pen an essay call *Il Saggiatore* (The Assayer). Galileo argued that comets were made visible by sunlight refracted off thin vapors high above us, rather than shining by their own light as stars do.

A DIRTY SNOWBALL

The argument over the comets of 1618 shows how over thousands of years our perception about cometary apparitions has changed. The oldest recorded comet came by around 1059 B.C. at the height of a battle between two ancient Chinese kings, Wu-Wang and Chou. Its eastward-pointing tail dominated the evening sky.

We've come a long way since then. By linking the appearances of comets in 1531, 1607, and 1682, Edmond Halley showed that comets orbit the Sun like the planets do. Since then, we have understood comets to be members of the solar system. Halley predicted that his comet would return in 1758. As the time of reckoning approached, mathematicians throughout Europe plotted out where the comet might appear. On Christmas night that year, a German farmer named Johann Georg Palitzsch trained his telescope at the right area of sky and sighted the prodigal comet.

The most famous comet of all, Halley has brought generations together since its first recorded appearance in 240 B.C. This comet takes a leisurely course through the solar system, moving outward from a point just inside the orbit of Venus all the way past Saturn, Uranus, and Neptune. In fact, Halley spends almost half of its time in the icy cold beyond Neptune's orbit. Its appearance at the height of the Norman conquest of England terrified both armies, and the Bayeux Tapestry proclamation *Nova Stella, Novus Rex!*— New Star, New King—reveals how seriously the comet's appearance was taken.

The famous Italian painter Giotto had three encounters with Halley: in 12 B.C., around 1300, and in 1986. It is likely that he gazed upon the comet himself at the beginning of the fourteenth century as he toiled on his "Adoration of the Magi." He could not have known that, instead of a star in the east, he was painting a comet that would be named for Halley. This comet, which rounded the Sun in 12 B.C., was the same comet he saw more than a millennium later. And in 1986, a spacecraft named Giotto in honor of the famous painting flew within 2700 kilometers of the potato-shaped nucleus of Halley's comet.

From the ground, Halley's 1986 return was not nearly so magnificent as its previous visit in 1910. In that year Halley cruised so close to the Earth that the outer reaches of its mighty tail brushed us. (Other than some reports of bright skies, the encounter had no noticeable effects.) Meanwhile a passel of unscrupulous entrepreneurs sold pills to ward off infection from the comet's noxious gases. The discovery of hydrogen cyanide gas in comets had undoubtedly spurred fears about comet contact. Although some comets do contain this substance, the amount of this material in a comet tail isn't enough to poison anyone.

Halley should be a little more impressive in 2061 than it was in 1986. As in that most recent encounter, Earth will be on the wrong side of its orbit when Halley reaches its closest point to the Sun, or perihelion. The comet will be as bright as always, but we will be too far away to see it to best advantage. However, in 2134 we will have to duck. There is no chance of an impact, but the comet will whiz by us in its closest approach to Earth since 837.

Although scientists understood orbits in 1758, they were nowhere close to figuring out the nature of comets. That insight would emerge in 1950, when Harvard's Fred Whipple suggested that comets were conglomerates of ices and rock—essentially dirty snowballs. This model has survived to this day, with one minor modification: The possibility exists that comets have so much more dust than ices that perhaps they should be called icy mudballs.

Throughout most of its life, a typical comet is a small nucleus several kilometers across. But in the few short months during which a comet nears the Sun, it changes its appearance dramatically. In the heat of the Sun's rays, the ices sublimate, or turn directly from solids to gas, and from gas to ionized gas or plasma. With the loosening of the comet's icy glue, dust starts to fly away from the nucleus with the outflowing gas. Within a few weeks, or perhaps even just a few days, the comet swells to an enormous size as its tiny nucleus becomes enveloped in a coma at least the size of the Earth.

Still approaching the inner part of the solar system, the comet faces the pressure of light racing from the Sun. Its coma material is now pushed away from the Sun, forming a tail. Almost all comets have dust tails, which can appear as fans or as long, curving streaks. But a comet that is actively shedding its ices may also develop an ion tail, also known as a gas tail, as gas storms off the nucleus and becomes ionized by ultraviolet radiation from the Sun. This tail can change its appearance radically within minutes as the ions are dragged by the solar wind. The ion tail points straight out behind the comet, while the dust tail, with its more slowly moving particles, is usually slightly curved.

Large comets that approach close to the Sun can produce spectacular pyrotechnics. As weak areas of their nuclei rotate into full sunlight, they tend to erupt, sending large jets of material into the comet's sky. In the fall of 1985, I used a telescope equipped with a CCD to record many images of jets firing off Halley's nucleus.

THE ORIGIN OF COMETS

Most scientists think that comets formed very early in the solar system's development. As an enormous, rotating cloud, the infant solar system had a huge concentration of material at its center, while small, solid bodies were condensing in eddies at various distances from the center. Most of these small bodies were comets. Billions of them were flying throughout the system. Some accumulated to form the outer planets, but most only flew close to growing planets, changing their orbits as they brushed past. These close encounters happened often in the solar system's early history, and many of them involved Jupiter, whose enormous gravity generally would fling a comet out of the solar system. Comets orbiting in the outer reaches between the newly forming planets Uranus and Neptune—and just beyond Neptune, on the other hand—were sometimes flung out to great distances but did not quite escape. The tidal forces in the galaxy and perturbations by passing stars then spread these comets out into a great spherical cloud that now surrounds the planetary system at distances ranging from a few thousand to nearly a hundred thousand times the distance between Earth and Sun, an area known as the Oort cloud.

Thus, the region of the planets was slowly cleared of its hoard of small bodies. But in the process of clearing, comets pelted the inner planets during the period that began with the ignition of the Sun about 4.5 billion years ago to about 3.3 billion years ago. One theory holds that this period of heavy bombardment gradually died down about 3.3 billion years ago. In a different scenario, the early bombardment leveled off around 4.3 billion years in the past to a long period of relative calm. Then a pulse of material, possibly from an object that broke apart in the inner solar system between Mercury and Mars, created an event we call "late heavy bombard-

Astronomer and artist William K. Hartmann portrays this view of the world of a comet nucleus. Courtesy W. K. Hartmann.

ment," which peaked around 3.9 billion years ago. The bombardment never died away completely, and comets continued to hit planets throughout geologic history. It is mind-boggling to think that humanity would actually witness such a profoundly fundamental event in July of 1994.

GREAT COMETS OF THE PAST

Until 1993, a great comet was synonymous with a bright one. Possibly the brightest comet ever to banner the sky came by in 1402. From March 22 to 29 the long-tailed exclamation mark was visible in broad daylight.[3]

On March 25, 1811 (exactly 182 years before the discovery of Shoemaker–Levy 9), Honoré Flaugergues found a fifth-magnitude comet. On April 11 the French comet hunter Jean-Louis Pons also found the comet. By December, it had a tail more than 70 degrees long, covering much of the sky. The coincidentally fine wines from that year were attributed to this comet. I also believe that John Keats was thinking of it when he penned his famous sonnet "On First Looking into Chapman's Homer" only three years later:

Much have I travelled in the realms of gold,
And many goodly states and kingdoms seen;
Round many western islands have I been
Which bards in fealty to Apollo hold.

Oft of one wide expanse had I been told
That deep-brow'd Homer ruled as his demesne;
Yet did I never breathe its pure serene
Til I heard Chapman speak out loud and bold:

Then felt I like some watcher of the skies
When a new planet swims into his ken;
Or like stout Cortez when with eagle eyes
He stared at the Pacific—and all his men
Looked at each other with a wild surmise—
Silent, upon a peak in Darien.

The middle years of the nineteenth century produced some magnificent comets. The great comet of 1843 kept itself pretty well hidden until the day it rounded the Sun. On February 27 it grazed to within 900,000 kilometers of the Sun's photosphere. During late March it put on quite a show, its tail brightening the sky south of Orion.

A comet found in Cassiopeia brightened into the Great Comet of 1854. For a couple of weeks in late March, it shone at a brilliant zero magnitude—as bright as the star Vega—with a short, bright tail. Just four years later, a splendid comet attracted wide attention. Discovered from Florence on June 2, 1858, Donati's comet brightened spectacularly. By October 9, when the comet passed closest to the Earth, it sported a tail some 35 degrees long—covering more sky than the full length of the Big Dipper.

Finally, in 1861, John Tebbutt, an experienced amateur astronomer from Australia's New South Wales, discovered a faint patch of cometary light with his unaided eye. Heading directly north, the comet brightened quickly. By the time it reached perihelion on June 11 it had swung behind the Sun and was completely invisible.

Around June 29 Tebbutt's comet suddenly reappeared in the morning sky as it pulled away from the Sun. It was now visible in the northern hemisphere, its head near the bright star Capella and its tail covering over half the sky, stretching to the constellation Hercules. According to some calculations, the Earth passed through the outer reaches of the tail on June 30. Although some people reported strange sunsets and a yellow tinge to the sky, the Earth survived its brush with this Victorian era traveller.

TWO GREAT COMETS IN ONE YEAR

The 1880s saw a second comet charge, which began one week after New York's Dudley Observatory was reopened following a period of new construction. One of the dignitaries at the ceremony casually remarked to Lewis Boss, the director, that people at other observatories were finding new comets. Boss turned to his assistant, C. S. Wells. "You see, Mr. Wells," he smiled, "you must discover a comet."[4]

Boss smiled again just one week later on learning that his enthusiastic young assistant actually met his challenge by discovering a comet. Comet Wells was always an impressive sight, even, as Boss noted, when it was far from the Sun and the Earth and looked like a big comet in miniature. But by June 7, the comet was visible in the bright sky just 10 minutes after sunset, and two days later it was visible even at noon.

Comet Wells had barely enough time to fade when sailors on an Italian ship noticed another new comet on September 1, 1882. Within five days the sun-grazing comet's head was as bright as Venus, and it sported a very long tail. By September 30, the comet's nucleus started to show some very interesting behavior. It split into two, then four separate comets, each on its own orbit, each to return hundreds of years in the future.

THE STRANGE CASE OF BIELA'S COMET: CAN COMETS JUST FALL APART?

When Jacques Montaigne discovered a comet from Limoges, France, in 1772, he had no idea that he was opening the book on a very interesting story. His comet faded after about a month, and when Jean-Louis Pons found it again in November 1805, the new comet was suspected of being a return of the earlier one. On February 27, 1826, Wilhelm von Biela discovered the comet for a third time, and also linked it to the two earlier comets. The acid test came on its next return in 1832, when the famous English astronomer John Herschel recovered it as it brightened on its way to the inner solar system.

Herschel's observation clinched this comet as only the third one known to be periodic, the other two being Halley and a small comet that whips about the Sun every 3.3 years and bears the name of Johann Franz Encke, the astronomer who unlocked its periodic secret. (We define a periodic comet as one orbiting the Sun in fewer than 200 years.) Instead of orbiting the Sun in a wide curve that would bring it through the inner solar system once in hundreds or thousands of years, Biela's comet returned every six years. Changed by repeated encounters with large planets like Jupiter, the comet's orbit had tightened over many years to its short ellipse.

But since it was not well placed in the sky, no one saw Biela's comet when it returned in 1839. On November 26, 1845, de Vico found the comet on its next return, and observations showed a normal comet brightening over the following two months. On January 13, 1846, the U.S. Naval Observatory's Matthew Fontaine Maury was conducting a routine observation when he noticed that the comet had two separate condensations—it had divided in two. Since the comet was still approaching both Sun and Earth, it brightened quickly, and its two components started to move apart. As it turned out, the comet had split at least two years earlier, but the two parts were not detectable before 1846 because they were too faint. Astronomers anxiously awaited the next return in 1852; on August 26 the brighter component was recovered, and the fainter one a month later. The comets' unfavorable return in 1859 was not detected. Although the visit in late 1865 was supposed to be a good one, with the comets prominently placed in the sky, diligent searches failed to reveal any trace of either comet.

Where were they? Had the comets simply disintegrated? A probable answer came in 1872, when they were expected at yet another favorable return. This time, in fact, the comet was predicted to come fairly close to Earth in late November. Again, searches revealed nothing. But then came the night of November 27, when people all over the world saw a major storm of more than 3000 meteors per hour. In 1885 there was an even stronger storm, with some 15,000 meteors in an hour. Another major storm occurred in 1892 and a smaller shower in 1899. Captured by Earth's gravity, the dust and rock that once comprised Comet Biela were raining down in fiery streaks. Since then there has been no evidence of Periodic Comet Biela—either as a comet or as a shower of meteors.[5]

In this century, other astronomers have tried to recover what might be left of Periodic Comet Biela. In 1971, for example, Lubos Kohoutek, a well-known astronomer, attempted to find Biela's comet using a large telescope. Nothing turned up except two new asteroids called 1971 UP1 and 1971 UG.[6] It is possible that one or two tiny cores a few hundred meters across still exist. If it exists at all, the comet is probably too faint to be observed. Comets do disintegrate from time to time: In 1993 Jean Mueller discovered her ninth comet. Although it brightened so that it could be seen with small telescopes, it later fell apart into a diffuse mass, becoming invisible to all but the most powerful telescopes.

RECENT BRIGHT COMETS

Although bright comets have been rare in recent decades, there have been a few fine ones. Comet Bennett in 1970 shone like a torch in the eastern sky. With its long tail, the apparition confused some Egyptians fighting a war of attrition against Israel into thinking that it might be an Israeli missile.

Six years later, Richard West discovered a comet on photographic plates exposed with a Schmidt camera at La Silla, Chile. Checking plates taken five months earlier, West was then able to find prediscovery images of this same comet. Using accurate positions from these plates, Brian Marsden predicted that Comet West could become a bright naked eye object as it got close to the Sun in early March 1976. The comet exceeded expectations. Brightening to magnitude minus 1 (almost as bright as the planet Jupiter), it painted a glorious picture in the predawn sky. As it came close to the Sun, Comet West treated viewers to a grand finale by splitting into four pieces.

Comet West was very bright, but the press all but ignored it. Part of the reason was that the comet was bright only in the predawn sky, as opposed to the more convenient evening sky, but the real problem was that everyone felt bitten by an episode that had occurred only three years earlier. It had to do with another comet that had shown some promise of putting on a super display.

COMET KOHOUTEK

Of all the bright comets this century, the most challenging one appeared in 1973. The story began when Lubos Kohoutek was trying to recover the same two asteroids, 1971 UP1 and 1971 UG, that he had found in his attempt to recover Biela's comet in 1971. While searching for 1971 UP1, Kohoutek stumbled upon a comet that, while never bright enough to be seen through small telescopes, displayed an unusual two-staged coma—an inner, bright cloud surrounded by an outer one that was much fainter.[7] Less than three weeks later, now looking for 1971 UG, Kohoutek found a second comet as faint as the first. This comet was different. Way out at Jupiter's distance at discovery, it was headed toward a close approach to the Sun nine months later. Once enough positions were known, Brian Marsden suggested that it could become as

In the summer of 1990, Comet Levy put on quite a performance as it moved through the sky. These five pictures show how the comet changed its shape as its tails, one made of dust, the other of ionized gas, grew in length. Photos by the author, using the 18-inch telescope at Palomar Mountain.

bright as magnitude minus 10—as bright as the Moon at gibbous phase. Even Harvard's venerated Fred Whipple dared to suggest that Comet Kohoutek could be the brightest comet of the century.

Excitement grew rapidly as this comet closed in on the Sun. The evening news eagerly tracked its progress, and users of small telescopes arose in the predawn hours to spot it. But two things happened just before Christmas 1973. As it neared perihelion, Comet Kohoutek's steady rise in brightness leveled off. Because it was a comet possibly on its first visit in from the Oort Cloud, its initial brightness surge was misleading, giving astronomers the impression that its bite was as big as its bark. It wasn't. The comet did brighten to magnitude minus 3—almost as bright as Venus. But on that day, December 28, it was so close to the Sun that only the astronauts aboard the Skylab space station had a good view of it. By the time it was far enough from the Sun to be generally visible, the comet had faded to a disappointing fourth or fifth magnitude—barely visible to the naked eye. "The Comet of the

Century," the Montreal *Gazette* editorialized, might just slip by without being seen.

If the comet was having trouble being seen, it was certainly having no difficulty attracting attention. Cruises to warm climes were arranged to give comet- and sun-bound passengers a chance to observe the apparition. And a religious cult called the Children of God passed out literature announcing that the brilliant comet was a dire warning of the end of life on Earth. Standing on street corners all over the world, its followers handed out sheets of paper detailing the consequences the comet would bring down on our vice-ridden society.

To most people, Comet Kohoutek was a dud. But the comet bombed only because our expectations for its performance were so unduly high. Kohoutek was actually a fine comet. On the evening of January 6, I drove with my sister-in-law's father, Irv Levi, to a dark site off a highway west of Montreal. The night was bitterly cold, the temperature below zero Fahrenheit. We set up a small telescope by the road, and I quickly located the prodigal comet. Though a little fainter than predicted, the comet was easy to see with its bright coma and a short but sweeping tail.

Our peaceful session was suddenly interrupted by a policeman whose car's flashing red lights added a little zest to our telescopic view. He wanted to know what our strange contraption was. All we had to do was mention the comet and he rushed to see for himself. He peered through the telescope for a minute or so. Then he looked us over. Finally shaking his head, he jumped back into his warm car and rushed away. No doubt he harbored a suspicion that we were foreign agents following some sort of missile.

We were both frozen, and I had a ticket to that evening's Bob Dylan concert. Back to the city we went. As I listened to Dylan wail "Like a Rolling Stone," I thought of our crazy hour watching that distant gem twist and turn its way through space. Some might have thought it a bust, but I'll never forget that frigid night in January 1973, when Dylan sang of a rolling stone on Earth, and I saw one in the sky.

The Planet King

It was prescient of the ancient Greeks to attach the name of their sovereign god, Jupiter, to that particular planet. They could have had no idea just how giant a planet Jupiter really is. More than 92 percent of all the material in the solar system, except the Sun, is part of Jupiter and Saturn. Jupiter has 318 times the mass of the Earth, and 11 times its diameter. We pause now to learn about this central figure in our comet crash story.

GALILEO AND THE MOONS

In 1609 an Italian scientist named Galileo Galilei heard of an astonishing invention from Holland. It was a device that consisted of two lenses placed a certain distance apart. By looking through the two glass lenses, one could magnify distant objects. Although history doesn't credit Galileo with the invention of the telescope—that honor probably goes to the Dutch spectacle maker Hans Lippershey—Galileo was ingenious in recognizing its awesome potential. Increasing its power to 10, Galileo turned his telescope toward the sky and observed the brightest objects—the Moon, Venus, and Jupiter. Discoveries like craters on the Moon, the phases of Venus, and spots on the Sun came quickly. But by far his most stunning find, and one of the greatest discoveries in the history of science, was a set of three stars near Jupiter. The following night Galileo saw the three stars in different positions, and a fourth appeared. From these sightings Galileo concluded that Jupiter had four moons. For the first time, worlds had been found that clearly did not revolve about the Earth.

To anyone who looked through Galileo's telescope, it was obvious that Ptolemy's system, which held that everything must revolve about the Earth, was no longer viable. In just a few nights of observation, Galileo could see the moons circling Jupiter. It was Galileo's greatest moment. It was also the beginning of his downfall.

The year after Galileo's discovery of the moons, the Catholic Church filed the first of several documents against him—in secret and apparently without his knowledge. In 1616, the codex against Galileo announced that the Church would specifically forbid any teaching that the Sun is in the center of heaven. Galileo decided to wait until a more propitious time to defend the Copernican view that the Earth revolved about the Sun. In 1623, the election of Pope Urban VIII, a modern thinker who seemed open to new ideas, convinced Galileo that his time had come. In the first years of the new papacy Galileo walked and talked with the pontiff. But the aging scientist had more enthusiasm than sense, since the Pope saw Galileo's observations as challenging the authority of the Church. In 1632 Galileo published his *Dialogue on the Great World Systems*, in which he clearly put forth the view of his telescope in the form of a debate. Imprudently, he created a character named Simplicio, who parodied the Pope's ideas.

Urban VIII was enraged. He felt that Galileo had taken advantage of his friendship, that through Simplicio he was poking fun at the papacy. Without further fanfare the Pope put the old and virtually blind astronomer at the mercy of the Holy Office of the Inquisition. Threatened with torture by being shown the instruments thus employed, Galileo was forced to "abandon the false opinion that the sun is the center of the world."

Compelled to live out his days under house arrest, Galileo never knew how important a tool his telescope would become. He would never know that a giant version of it would someday orbit the Earth, and that a spacecraft honoring his name would visit the moons he discovered.

JUPITER THROUGH THE CENTURIES

Our understanding of the mighty planet Jupiter has grown over the centuries since Galileo first examined it through his spyglass in 1610. Decades later, Isaac Newton calculated the orbits of Galileo's four moons, and then used them to estimate Jupiter's mass and density. With the development of bigger and better telescopes in the 1700s, observers began to notice the planet's markings. What we know now, of course, is that they tell us only what is going on at the top of Jupiter's troposphere; the inner layers remain hidden from our view. By the nineteenth century, observers could track the Jovian features with their large refractors. The changing appearance and motion of the markings allowed these observers to track Jupiter's strong atmospheric winds.

By the 1930s, observers were starting to detect methane (CH_4) and ammonia (NH_3) in Jupiter's atmosphere. Molecular hydrogen (H_2) turned up in observations after 1960. With the advent of high-quality, infrared telescopes in the 1970s, observers added ethane, acetylene, hydrogen cyanide, carbon monoxide, and other materials to the witches' brew that makes up Jupiter. These substances are minor constituents of an atmosphere that is mostly hydrogen, with a subordinate amount of helium.

ABOUT JUPITER

Although Jupiter shines by reflected sunlight, it emits more energy than it receives from the Sun. This is the energy it stored up while it was contracting from a much larger nebula of gas near the dawn of the solar system some 4.5 billion years ago. The nebula was not formed at the very dawn of the solar system; it formed around an older planetary core.

In fact, Jupiter still radiates about twice as much heat as it absorbs from the Sun. It was a larger and hotter planet at its birth 4.5 billion years ago, when the solar system was new, but it shrank and cooled. It is still emitting that primordial energy as it contracts. Although Jupiter releases so much energy, it is a very long way from becoming a star. It is not a sun because nuclear fusion of hydrogen to helium does not take place inside. Had it been some 50 times more massive than it is now, its inside temperature would have been high enough to begin nuclear fusion, and our solar system would have been ruled by a double star. It also has very strong winds—gales blowing at several hundred kilometers per hour—in its upper atmosphere.[1]

Jupiter is so large that it almost represents a transitional stage between planets like the Earth and a small star. The 16 moons that whirl about it look like a miniature solar system, with Jupiter as the Sun and the moons as planets. Two of its moons, Ganymede and Callisto, are bigger than the planet Mercury.

THE RED SPOT

Churning its way through the tops of Jupiter's clouds is a storm at least as wide as the Earth and three times as long. It has survived for at least 300 years. Without some constant engine to power it, this giant scarlet-hued hurricane might not have lasted this long. It is possible that the spot, along with other long-lived storms, is a large convective region where gases are condensing inside Jupiter. As the gas condenses, it releases heat, providing energy to keep the spot active.

The Great Red Spot is generally harder to see now than it was when I first started observing the planet in the 1960s. Back then the spot was markedly redder than the surrounding material. By the mid 1970s it had lost some of that contrast, and I still often have difficulty seeing it. It does not appear that the storm itself has subsided, since it is still the same size as it was.

VOYAGER

March 5, 1979: Almost four centuries after Galileo's trial, a room-sized spacecraft sashayed by Jupiter and took pictures as beautiful as Rembrandt paintings. It was not the first visit to the outer solar system by a spacecraft. Pioneers 10 and 11 passed Jupiter in 1973 and 1974 and showed a cloud deck far more complex than Earth-based telescopes had revealed. But Voyager's cameras brought back not a fast look, but a series of gorgeous pictures showing, for example, the Great Red Spot actually rotating.

One reason the two Voyagers did so well was that they started photographing Jupiter two months before their encounters with the big planet. Every two hours they snapped a detailed series of images of Jupiter in different color filters. As the crafts rushed by, the images they recorded covered smaller sections of the planet, so the best pictures were taken while the craft was not at its closest to Jupiter.[2] However, when the craft moved to the night side, one surprise came after another. Voyager recorded Jupiter's aurora, a meteor, and even lightning, all beautifully lit against the darkened planet.

As anyone who has seen Jupiter through a small telescope can tell, the giant planet's atmosphere has very rich colors. (The colors shown on Voyager photographs are "false color" computerized images, exaggerated from the real colors.) On Jupiter, sulfur, phosphorus, and carbonaceous materials may tint the features. Moreover, color in clouds appears to indicate their altitude. The lowest clouds are blue and are seen only through breaks in those above them. Brown, and then white, clouds are higher up, with red clouds like the Great Red Spot at the loftiest levels.

The Voyagers enabled our understanding of Jupiter's clouds

to mature in 1979. Instead of dozens of planetary markings, there were now thousands of features, most of them temporary, whose positions and movements across the planet could be measured precisely. Since we knew that helium had been present in the primordial solar system, we knew that this element should exist in Jupiter. By detecting the results of collisions between hydrogen and helium molecules, the spacecraft indeed made the technically difficult discovery of helium on Jupiter.

Jupiter's cloud layers seem to be organized into at least three zones. Hypothetically, at the top is an ammonia ice cloud, beneath which is a layer of frozen crystals of ammonium hydrosulfide, actually a tossed salad of ammonia and hydrogen sulfide. The lowest layer is water, in the form of liquid drops or ice crystals. As anyone can see through a moderate-sized telescope, Jupiter's atmosphere is organized into bright zones and dark belts. In each belt, high-speed winds blow opposite each other from the east and west. The Earth also has such counterflowing winds. At midlatitudes these winds, called the jet stream, flow from the west, while the easterly trade winds blow in tropical latitudes. On Jupiter, several winds blow in opposite directions on each hemisphere. These "zonal jets" have remained at much the same latitude over almost a century.[3]

Major storms, which we call vortices, often form along the boundaries of zonal winds flowing in opposite directions. Unless they spin around as quickly as the velocities of the winds, these storms usually don't last much more than two weeks. The speed at which a vortex spins determines how long it will last. If it spins too slowly, the zonal winds around it will stretch it and then annihilate it. Large storms can become quite stable by whirling with the zonal winds. The Great Red Spot, itself a giant Jovian vortex, is so secure that the Voyager spacecraft observed smaller clouds that come up against it from the east and then took about a week to meander around it.[4]

Why is there a difference between wind features on Earth and on Jupiter? One explanation: On Earth, the temperature difference between the equator and the poles helps drive Earth's midlatitude cyclonic flows of air. On Jupiter, there is little difference in temperature from the equator to the poles. For the temperature to be so uniform, it would seem that energy has to be transferred from deep in the interior to the surface.

This is why Comet Shoemaker–Levy 9 presented such a marvelous opportunity. Each of the comet fragments is a space probe that penetrates Jupiter's clouds. Galileo is a different kind of probe, in that the spacecraft's penetrator device has a set of detection instruments and a communications system to report their data, via an orbiting spacecraft, to Earth. But by the way the comets behave as they slalom into Jupiter, they are capable of telling us a lot of things, if we can only understand their language. For example, if a lot of water were to surge upward, we could conclude that the comet penetrated the water later, and then deduce the abundance of water and other materials at that layer. And if the impacting comet sends a seismic wave through Jupiter, we might learn about the size of the planet's core. Confirming what we suspect about Jupiter's interior would be an impressive reward for our studies of the impact: Beneath the cloud deck is an envelope of hydrogen and helium, a large spherical layer of metallic hydrogen, and at the middle, possibly a solid core about the size of the Earth.

JUPITER'S RING

Until 1979, most astronomers assumed Saturn was the only planet decorated with rings. Although few expected to see a ring when Voyager 1 visited, the craft did take an 11-minute-long time exposure—just in case—while the spacecraft was in the shadow of Jupiter. The 11 minutes were well worth it. Jupiter does have a thin ring; although it is faint, it showed up clearly in the Voyager images and has since been followed up through ground-based telescopes. Using Palomar's 200-inch telescope, Phil Nicholson was the first to detect Jupiter's rings from Earth. Using a small device called a coronagraph to block the light of Jupiter, the Lunar and Planetary Lab's Steve Larson was able to photograph the ring with the 61-inch Catalina telescope near Tucson. The ring's outer boundary is almost a Jupiter-diameter from the tops of the planet's clouds.

Jupiter's thin ring is made up of minuscule particles of rock or ice probably no more than a few microns across. It would not last very long—perhaps only a thousand years—were it not for two of Jupiter's moons, Metis and Adrastea. They orbit near the ring, keeping it in place and supplying particles to it. Jupiter may also have a sizable herd of boulder-sized microsatellites that contribute material to this ring.

JUPITER'S MAGNETOSPHERE

The Sun's outer atmosphere, or corona, continuously vents intensely hot, ionized gas in the form of a solar wind. Although this wind does not influence the movements of planets in their orbits about the Sun, it does have two important effects: It allows the ionized gas tails of comets to develop and point away from the Sun, and it contributes charged particles to regions around the planets we call magnetospheres.

We define a magnetosphere as the space where a planet's magnetic field controls the patterns of ionized gas. The Earth's magnetosphere is strongest near the poles, where charged particles occasionally interact with it and form the aurora borealis and aurora australis. In 1958, James Van Allen used the radiation detectors aboard the first American artificial satellites, Explorers 1 and 3, to discover the energetic particles trapped in the Earth's magnetosphere. Two different areas, since called the Van Allen belts, surround the Earth.[5] On our planet's sunward side, the magnetosphere extends out to a few Earth-diameters until it meets the bow shock formed by the incoming solar wind. In the other direction it extends much farther out into a tail.

Our magnetosphere is nothing compared to Jupiter's. Extending as far as 25 Jovian-diameters (actually, astronomers prefer to say 50 Jovian radii), it measures some 1200 times bigger than that of the Earth. If we could see it, it would appear as the largest object in the solar system. It surrounds Jupiter to a distance of more than a degree, or two Moon-diameters, on either side as viewed from the Earth. Away from the Sun, the magnetosphere's tailward side stretches for several times the distance between Earth and Sun.

OBSERVING JUPITER

Whether Jupiter is being battered by comets or not, it is always a joy to look at. Even through nothing more powerful than a good pair of binoculars, Jupiter's four Galilean moons should be visible, their positions changing noticeably from one night to the next. The smallest telescope reveals features on Jupiter's cloud tops, including two dark bands straddling the equator. Through larger telescopes, other dark belts and bright zones appear, as well as exciting detail within the belts.

The best way to learn about Jupiter through observation is to draw it. Observers use a soft, 2B pencil and a dim white flashlight so that they can see what they are committing to paper. Before beginning to draw, they watch the planet for a few minutes to get familiar with the shapes and details of its belts and zones. Since Jupiter rotates very quickly—the whole planet goes once around in less than 10 hours—observers complete the basic outline of their drawings in about a quarter hour, filling in the details later.

The experience of drawing this planet brings to mind the fact that Jupiter is big. It is a planet much larger than Earth and some 400 to 600 million miles from us. While you look at Jupiter's moons, consider how they helped persuade Galileo that the Earth was not the center of the universe, and remember that the idea was so threatening to that era's powerful religious politics that he was forced to recant on pain of torture. By taking us back to an earlier, darker time in our history, Galileo's moons remind us not to be too attached to the accepted wisdom of our own age.

6

Preparations Begin

Astronomers will organize to coordinate studies of the probable demise of what already is surely the most interesting comet ever discovered.[1]
—Clark Chapman, June 1993

"We're going to see an impact!" Gene Shoemaker said on May 22, 1993, when Marsden published his epochal announcement. Scientists were thunderstruck, and within a few days three groups of scientists were getting involved. Comet people, already excited about viewing a disjointed comet so soon after breakup, relished the chance to learn a lot about how comets behave during those critical moments when they encounter a planet. Jupiter experts saw Shoemaker–Levy 9 as their golden key to unlocking some long-guarded secrets of that planet's interior. And impact physicists, never having seen an explosion anywhere close to what the speeding fragments of S–L 9 could provide, looked forward to studying in complete safety, without risking humanity's destruction, the releases of energy hundreds or thousands of times stronger than all the world's nuclear arsenal put together.

Some early predictions were wildly enthusiastic. If each of the fragments were 10 kilometers, or almost 7 miles, wide, the scientists said, then their impacts could release the equivalent of a 200 million megatons of dynamite. Compared to Hiroshima's 20,000 kilotons, these explosions would be unimaginably huge. If the explosions took place on the side facing us, they said, Jupiter would light up like a firecracker, shining brighter than the full Moon for a minute with each impact and rendering Jupiter visible with the naked eye in full daylight. But with those first predictions came some disheartening news: According to orbit calculations, all the impacts would be on Jupiter's night side, not visible from Earth. Just how far behind the day side of Jupiter the fragments would strike was a matter of hot debate.

The difficulty lay in the fact that we still did not have a really good idea of the curious comet's orbit. Despite the logging of several hundred precise positions, observers had kept tabs on the comet only since March, and they were being asked to produce their measurements of the positions of each of the fragments. Brian Marsden's approach throughout this early period was to use only measurements of the center of the train, and not of each of the fragments. As long as the comet fragments were close together, this approach worked well, allowing predictions of the time and place of impact for the center of the entire train. Some researchers thought that this information was not accurate enough and that observers should be sending in accurate measurements of every nucleus of the train. But Brian did the best with what he had. "Almost all the observations," he said, "were from observers whose telescopes were not powerful enough to get good images of each fragment."[2] If he had not used these observations, there would simply not have been enough observations to work with.

Marsden kept his strategy of using midtrain measurements until November 19, when Jim Scotti and his student Travis Metcalfe submitted accurate positions for nine of the nuclei over several months. On November 22, exactly six months after his announcement of a pending collision, Marsden was able to write that it was now absolutely certain that all the comets would collide with Jupiter.

Initially it was hard just to know which fragment was which. With the first publications about the comet, two competing systems were adopted. From the Jet Propulsion Laboratory, Zdenek Sekanina's scheme used letters of the alphabet to identify each of the 21 fragments, east to west from A to W. This was a traditional system used for the fragments of Periodic Comet Brooks 2, in 1889, and for Comet West, which split into several pieces in 1976. But David Jewitt of the University of Hawaii proposed a different approach, with pieces numbered west to east from 1 to 21. Since the two schemes began in opposite directions, getting everything straight was a headache. Thus we ended up with the unwieldy Q = 7, R = 6, and S = 5 for the various fragments.

THE PRECRASH BASH

By early July 1994 it had become clear that some meeting of the minds was needed both to figure out what was going on with S–L 9 and to start coordinating plans for observing the impacts. Held at the University of Arizona's Lunar and Planetary Laboratory under the 100-plus degree summer heat, the "precrash bash" was characterized, Jim Scotti reported, by "many arms waving about in all directions" as people tried to offer explanations based on the limited number of observations.

One basic question arose: What was the probability that the comet would get caught in an orbit about Jupiter? Gene Shoemaker suggested that the giant planet has one comet of about a kilometer in diameter captured in orbit about it at any given time. He noted that at least two other comets had been trapped in temporary orbits about Jupiter within the past quarter-century: Periodic Comets Gehrels 3 and Helin–Roman–Crockett. These orbits didn't last more than a few years before the comets returned to their travels directly around the Sun. But these two comets are much bigger than a kilometer in diameter—maybe they are larger than 4 kilometers—and the number of comets in orbit is thought to increase steeply with decreasing size.

A comet breaking apart after passing through Jupiter's Roche limit was another matter. Comets do not have to be in orbit about Jupiter to do that. Periodic Comet Brooks 2 was orbiting the Sun directly when it closed in on Jupiter and passed within the planet's Roche limit in 1886 and broke apart. For a comet to do both—break apart while in orbit about Jupiter—is a rarer event. Add to that the tiny chance that the comet continues to orbit Jupiter and then crashes into the planet, and we have something really singular.

Whatever the odds for all three occurrences—orbit about Jupiter, breakup, and collision—the meeting participants agreed that we were fantastically lucky to see all three happening before our eyes. The breakup had allowed the comet to brighten so that it could be discovered, and the orbit about Jupiter gave us the time to plan our observations. To combine these rare events with an impact that we could observe and study was fortunate in the extreme, a circumstance unlikely to occur more than once in a thousand years.

While Tucson's precrash bash was going on, NASA's Jürgen Rahe spiced things up dramatically by announcing that 750,000 dollars would be reallocated for impact studies. Morris Aizenman, of the National Science Foundation, was personally present at the meeting to say that NSF would ante up an equal amount of funding. Although these amounts were small by government budget standards, the decision to spend them indicated the unanimous agreement that the event deserved immediate attention. "We want to make it as easy as possible to accomplish what needs to be done," Aizenman quipped.

GETTING THE ORBIT RIGHT

By the time the comets skidded into the Sun's dazzle in late July, the Jet Propulsion Lab's Donald Yeomans and Paul Chodas figured that it had been looping around Jupiter since at least 1970. Also, in several of its two-year orbits, it passed dangerously close to that gaseous giant. When the comet finally broke apart in July 1992, they computed, the fragments followed an orbit that took them almost as far from Jupiter as Mercury's average distance from the Sun. After reaching this considerable distance from Jupiter, they slowly started to fall back toward the planet.

If the Yeomans–Chodas orbit was correct, all the fragments would have at least a 99 percent chance of crashing into Jupiter at their next closest approach, or perijove (*peri* for closest point, *jove* for Jupiter). But for all the excitement, this orbit seemed a bit of a letdown. The nuclei, as we have seen, would strike Jupiter on its night hemisphere, albeit toward the dawn side. From our earthly vantage point, we wouldn't be able to see them crash. The 21-odd icy chunks would hurtle in one at a time over a six-day period beginning about July 18, plunging into the planet's atmosphere near 44 degrees south latitude. Because of Jupiter's rotation, instead of landing on the same place, they'd create a belt of turmoil at various longitudes around the planet. If we could see all this, we would get a good show after all.

HOW LARGE WERE THE COMET FRAGMENTS?

Two factors determine the energy released by an impact. One is the velocity of the projectile, and the other is its mass. Early on, we knew that the impacting comet nuclei would be traveling at 60 kilometers per second, or about 135,000 miles per hour. But calculating their sizes proved a more daunting task.

"It is difficult to argue statistically," Brian Marsden noted, "but I think there are many small short-period comets in the region between Jupiter and Neptune and that a fair number are temporarily orbiting Jupiter at any time. The bulk of these objects are 1 kilometer across (about six-tenths of a mile) and less. [Comet Shoemaker–Levy 9] has done a strange thing by passing close enough to Jupiter to break up on one perijove and actually to collide with Jupiter on the next pass. Odds would tend to favor a small object doing this. I suppose there has to be some lower limit of size in order to break up in the general way observed. Perhaps this is even as low as 100 meters, although—as I say—I am still prepared to go up to 1 kilometer or so for the original comet. There is not a prayer of detecting these objects unless they break up like this comet has done."[3]

Two other groups attacked the question of size theoretically. I heard about the first one afternoon in September 1993 when Jim Scotti called me excitedly to his office to discuss some work that he and Jay Mellosh had been doing. "We understand when the comet broke up," he said, "and we also know how big it was before it broke up." They based their conclusions on the assumption that the comet shuddered and fell apart like a ball of wet mud precisely at the moment of its closest approach to Jupiter on July 7, 1992. Since the train's length was directly proportional to the size of the progenitor—that is, the original intact comet—they concluded that the progenitor comet could not have been larger than about 2.3 kilometers across. But Yeomans and Chodas had a different idea. The way the comet train was oriented, they thought that the comet could not have split apart at its closest passage to Jupiter. The presence of the large wings of dust on both sides of the comet train supported their contention. They suspected that the comet had been large enough to hold itself together by its own gravity for 1.4 hours after coming closest to Jupiter. If that indeed happened, the progenitor would have had to have been large—at least 9 kilometers in diameter, and its largest fragment might be some 4 kilometers wide.

Just how large the original comet had been remained a major question during the entire precrash period. Despite many attempts by observers at ground-based telescopes, scientists were not able to get a handle on how big the fragments were. However, the Hubble Space Telescope data did set firm upper limits to the sizes of the nuclei and of the progenitor, sizes very similar to the values reached by Yeomans and Chodas.

If we knew how large each nucleus was—say about 1 kilometer, or three-fifths of a mile in diameter—and if we knew what each nucleus was made of—say solid ice—it would be easy to calculate the energy the impact would release at 60 kilometers per second. It would be 10^{28} ergs of kinetic energy, the equivalent of 200,000 megatons of TNT. If the comet measured 5 kilometers, the explosion would be 125 times larger. The size question would remain unanswered for some time after the impacts occurred.

EARLY IMPACT PREDICTIONS

Even if collisions between comets and planets happened every day, astronomers still would have been uncertain about the observable effects of Shoemaker–Levy 9 on Jupiter, since there were no reliable size estimates for any of the fragments. But this didn't stop at least four teams from firing up powerful computers to run models simulating the blasts. Generally, the models suggested that the comets would take just a few seconds to plow below the visible cloud layers. In a model by Caltech's Thomas J. Ahrens, during each violent event a shock wave would heat the atmospheric gases along the entry route to tens of thousands of degrees, creating what Ahrens calls a "tunnel of fire."

As the remaining fragment pushed into the tunnel, it would slow down, break apart, and vaporize. Both the fragment and the planet's gases surrounding it would be heated to a plasma that would rise back through the tunnel of fire. When this fireball broke through the clouds a fraction of a minute later, it would look like an imitation Sun hundreds of kilometers across. The expanding cloud of gas would cool quickly, turning a dull red before fading.

Besides that early scenario, there were two other possibilities. In one, the fragile fragments would disintegrate relatively high in Jupiter's atmosphere, possibly even above the topmost cloud deck. In another, the nuclei would penetrate so deeply that almost all the energy would bleed off into the tunnel of fire, and little would erupt above the clouds.

PLANNING THE ARMADA

Regardless of how energetic the impacts proved to be, astronomers were delighted to have more than a year's advance notice of a major planetary collision. Most observatories allocate their valuable telescope time months in advance and reluctantly change their schedules when major events, like the explosions of dying stars we call supernovae, force them to make last-minute changes in the schedules. The May 22 announcement of impacts, 14 months ahead, was a gift from heaven. Virtually every major observatory on our planet, and in space around it, was getting into the act. The list of facilities granting time for the big crash read like a Who's Who of observational astronomy, including big guns like Palomar's 5-meter Hale telescope. Kitt Peak National Observatory, usually closed during Arizona's summer monsoon season, made an exception to keep its two largest telescopes open for impact week. The giant infrared telescope on Mauna Kea planned to devote observing time to the event. Based at the National Aeronautics and Space Administration's Ames Research Center in California, the Kuiper Airborne Observatory had been scheduled for inspection and overhaul for both the aircraft and its 36-inch telescope. But since its infrared detectors could be crucial in measuring infrared spectral features, especially those of water vapor, this downtime was delayed until September. Never in the history of astronomy had this much telescope time been allocated for a single event.

At first, the roles for Galileo and Voyager 2 were unclear. Although still 18 months and 240 million kilometers from Jupiter, Galileo would be at the considerable angle of 50 degrees to one side of the Jupiter–Sun line. From this perspective, mission planners thought that the impacts would take place just a small distance beyond the planet's dark limb. Thus, Galileo might be able to monitor both the entry meteors and the tops of fireballs arcing high above the Jovian clouds.

Could the far-off camera of Voyager 2, shut off since 1990, be resurrected for the S–L 9 crashes? More than 6 billion kilometers from Jupiter, the spacecraft would barely see Jupiter as a point of light. Still, what Voyager might lack in resolution it could make up for in position. The predicted impact sites would be in full line of sight.

Two major problems haunted Voyager and Galileo planners. By the end of July, the uncertainty in the Yeomans–Chodas orbit was about a day. As more observations showing accurate positions of the comet came in, astronomers expected the impact window to be refined to about an hour by early May and to within 20 minutes by a week before the crashes. In fact, claimed Yeomans, the uncertainty might still be nine minutes just six hours before the first impact. But there is a favorable catch to all this. The locations of individual nuclei with respect to one another were known much better than their positions relative to Jupiter. So if an observer or a spacecraft could time the first impact precisely, the remaining times could be calculated more accurately.

THE HUBBLE SPACE TELESCOPE

Because it is in orbit, the Hubble Space Telescope could follow the comet long after it had moved into the bright glare around Jupiter and became lost to most telescopes on the ground. A fully repaired Hubble telescope would resolve details on Jupiter as small as 230 kilometers, rendering small changes in the clouds easy to spot. Other orbiting observatories and even the Russian space station Mir were lining up for possible observation roles in the play.

About two weeks after we discovered S–L 9, the Space Telescope Science Institute's Harold Weaver telephoned me from Baltimore. He wanted to know some details about the comet's early appearances. So long before there was any idea of an impact, the Hubble Space Telescope was already assembling a team to observe this extraordinary comet. I told him that both Gene and Carolyn Shoemaker and Steve Larson had been talking about the possibilities of HST. Weaver quickly became the head of an S–L 9 science team planning to observe the comet.

On July 1, the Hubble Space Telescope turned its astigmatic

eye toward Comet Shoemaker–Levy 9. It recorded the same number of fragments—21—that ground-based observers had seen, but it showed more detail about each one. The most interesting example of this was the complex of fragments P and Q; Q seemed so elongated that we concluded that it had split into two pieces. The Hubble images offered tantalizing views that didn't quite tell us enough; we hoped that the repair mission at the end of 1993 would lead to clearer pictures. The July 1 images were supposed to be repeated a week later, but the Hubble suffered a minor computer anomaly. Whenever that happens, the spacecraft returns to its fetal position—solar panels to the Sun, antenna to Earth, all instruments shut off—until operators solve the problem and restart the telescope. Although HST was up and running soon after its hiccup, planners decided not to repeat the S–L 9 observation, partly because the comet was already too close to the position of the Sun for the telescope to point to it safely. There is a risk that the telescope could actually look toward the Sun and fry its detectors. We had hoped to have two Hubble sessions with the comet, the better to study its evolution over time. With the comet, and Jupiter, rushing to conjunction with the Sun, this would be our last chance to see it for at least several months.

Little did we know at that time that the single image we got would keep the entire science team very busy. Although the telescope's poor optics kept us from seeing the comet as clearly as we would have liked, they still gave us our most detailed picture. From the magnitudes of the comet fragments, Weaver proposed that the maximum size of the brightest piece, Fragment Q, appeared to be 4.3 kilometers. This was well in line with the Sekanina–Yeomans–Chodas estimate.

SUMMER SOLSTICE, 1993— CHILDREN AND THE 61-INCH

On the night of June 21 I joined a group of children who were observing at the 61-inch telescope high in the Catalina Mountains northeast of Tucson, Arizona. Although I had often observed at that telescope, I had never used the instrument under these particular circumstances. It was fantastic. Ages 9 to 12, the kids were huddled on the observing platform. They were with Donald McCarthy's Astronomy Camp, an annual project of the University of Arizona, and I've made a point of talking with eager youngsters at this camp each June.

Although I work with children and the sky often, this annual event is special to me. From 1985 to 1990, comet scientist Steve Larson and I used this telescope to study Halley's comet and dozens of other comets, and seeing the crowd of young people sitting using this giant telescope brings me back to my astronomical roots.

Huddled on the platform that sits just below the great telescope, the astronomy campers watched as I opened my favorite battered copy of Leslie Peltier's *Starlight Nights*. "All day long they sit there," I quoted Peltier's writing of his telescopes long ago, "without a sign of life about them. It is only when the stars come out that they begin to stir."

At this point I quietly turned a lever, and the dome's huge shutters surged to life and began to open. Silently the kids looked upward as a darkening sky was revealed. I read on: "Then like some snowy owl and owlet waking for a night of dark marauding, they spread apart their tight-closed wings, open up their big round eyes and peer about in all directions. The prey they seek is written in the skies, for these are my observatories."[4]

The group sat quietly as the shutters finally opened all the way. That night we looked at two exciting objects in the southwest. With bright zones and dark belts crossing its face, Jupiter was the first thing we watched. Then we moved the telescope a short distance away so that it focused on the comet, whose pieces were spread apart much more than they were when Carolyn, Gene, and I discovered it three months earlier. In fact, they had spread out to the diameter of Jupiter itself, and even though we were looking through a large telescope, the comets had faded enough so that they were pretty hard to see.

All over the world, visual observers had noticed the comet's gradual fading—a completely normal phenomenon. As the particles of dust surrounding the nuclei spread apart, their reflected light was spread out over a larger area, and they faded. But as comets dimmed, the excitement of their impending doom caused the rumor and humor mills to brighten.

Even in this enlightened age, catastrophic events have the power to bring out the soothsayers, and S–L 9's impending demise offered a grand target for them. Sister Sofia Richmond published the first of two full-page ads (the second appeared in February 1994) in the *London Times* and other British newspapers:

A WARNING ULTIMATUM FROM GOD
TO ALL GOVERNMENTS!

REDUCE THE CRIME EPIDEMIC
DESTROY ALL PORNOGRAPHY
STOP ALL WARS OR FACE GLOBAL EXTINCTION.

Almighty God the Creator of this Universe has ordained this forthcoming cosmic event to warn all nations of what will happen to them if His messages are ignored. All governments; Britain Europe USA etc. must obey God's ultimatum at top speed or face extinction by fireball from Jupiter.[5]

The collision was being sent as a celestial warning, she propounded, and unless certain things were done, Earth could be next.

The scientific community reacted in different ways to this announcement. In England, astro-popularizer Patrick Moore angrily criticized the ad as taking public understanding of the sky back centuries. But I thought it was a nice little bit of fun. The Dark Ages passed from view long ago. People could believe that such events were sent as cosmic warnings if they liked, and I would hope that most readers of the *Guardian*, like most astronomers, took such soothsayers with the humor and disdain that they deserved.

As recently as 1910, when people could buy comet pills to ward off the evil effects of Halley's comet, the idea of comets as omens was being taken somewhat seriously. Fear of Comet Halley drove one hapless man out of a window to his death. Most people took the extraterrestrial visit with the curiosity and pleasurable excitement reflected in such musical numbers as *Halley's Comet Rag*.

BOULDER, 1993

By early fall, it was time for scientists to meet again to compare notes and begin to formulate observing plans. The opportunity came at the end of October. Each year, a select section of the American Astronomical Society, the Division for Planetary Sciences, meets in a different place. The DPS is the world's largest gathering of people who study the other planets, in much the same way as geologists study the Earth.

Until October 1993, the only professional meeting about S–L 9 was the multidisciplinary Comet precrash bash in Tucson. Although that confab was successful, it was intended only as a quick first look at the possibilities of what could happen and as an assessment of the observing situation: Who was planning to observe what? What telescopes would be needed? How would the resources be allocated? By early fall, if we were no closer to finding out any of the answers, at least we had a better idea of the questions.

The first challenge we faced was just getting to the meeting. I joined the Shoemakers in Flagstaff on the preceding Saturday for Gene's retirement party. After 45 years, he was retiring from the U.S. Geological Survey. He had founded its astrogeology branch and headed it for many years. There were speeches, anecdotes, even a song. The party was a truly deserved tribute, and the several hundred people present all had a wonderful time despite the cold rain coming down. Of course, everyone who knew Gene understood that he was not really retiring at all, but changing his port of call from the U.S. Geological Survey to Arizona's Lowell Observatory. And even that came gradually—a year later he still had the same office at the Geological Survey and worked every bit as hard as before. The day after Gene's gala sendoff, we arrived in Boulder, Colorado, in rain so dense we could hardly see the road from the airport. We checked into our hotel and joined the other exhausted travelers. An inauspicious start to the meeting, that first day ended with drinks for a bunch of exhausted travelers.

The next afternoon was just as wet, and colder. Almost 800 people gathered in the hotel's largest meeting room. I didn't realize how large the crowd was until I turned around and saw people standing in the aisles. I asked Carolyn, who was sitting next to me, if she had ever thought we'd find a comet so many people wanted to hear about. Carolyn turned around and stared at the large crowd. Gene's keynote speech reviewed the discovery circumstances and outlined what we knew so far about the history of the comet's orbit. He closed with a summary of the problem: The impacts would be on the far side, out of view of any earthbound and Earth-orbiting telescopes.

In those early days, it was hard to be certain what role the Galileo spacecraft might play. Shortly after the first definite orbit was calculated at the end of May, mission planners realized that their spacecraft, with its view of part of the planet's far side, had a good chance of observing the impact sites directly at the time of the collisions. But as the orbit solutions changed slightly in the early fall of 1993, the sites seemed to move ever further away from the planet's terminator, its frontier dividing day and night. By the Boulder meeting, it appeared that the only thing Galileo might see would be the plumes rising from the explosion sites, and only if those plumes were high enough.

HIDE AND SEEK

I left Boulder with mixed emotions. On the negative side, there seemed to be a growing confusion about what the effects of the collisions might be. The early predictions of flashes that would brighten Jupiter enough to be clearly seen in broad daylight were out the window, but aside from that, opinion on what would happen was divided. Also, the Galileo spacecraft seemed positioned to get a tantalizing view of the impact sites—close but no cigar.

The positive side of the meeting was more esoteric. Passing through nearby Denver on the way home, I stopped at the old site of a hospital residence where I had lived 30 years ago as a patient. Walking around the now-deserted grounds of the Jewish National Home for Asthmatic Children, I recalled the many nights I had spent there observing as a child. I was very excited about sky-watching even back then. It was a difficult time, away from home, school, and friends, but it was an important time too. I recalled how in the spring of 1963, I would wake up in the dead of night, walk for about five minutes with my telescope to a place away from street lights, and try in vain to locate the planet Neptune. I was wheezing mildly after these unsuccessful efforts, so after each predawn session I would amble over to the hospital building for some medicine. After a week my doctor asked me why I needed these nightly puffs. I explained how I was having an awfully frustrating week not finding Neptune, but that since I was getting wheezy I would stop this search.

Dr. Chai wrote some notes on his pad, then looked at me. "David," he intoned, "I'm writing up an order for more medication; so if you need it, go get it. But I want you to keep on searching until you find Neptune. Never let your asthma stop you from doing what you really want to do."

I've never forgotten that lesson. Three decades later, as I wandered these deserted but once famous grounds, I wondered what all the asthmatic friends I made there were doing now and what they'd think of the sight of 800 people listening intently in nearby Boulder to reports of Periodic Comet Shoemaker–Levy 9.

Shoemaker–Levy 9 Returns

By the end of July, Jupiter and Shoemaker–Levy 9 were setting earlier and earlier each evening until they were no longer seen in the night sky at all. During the intermission that followed, the comet fragments began to speed up as they left apojove, or the farthest point of their orbit about Jupiter. On July 16, 1993, they started creeping back toward Jupiter from a distance of 50 million kilometers, at a speed that would increase very slowly until the last weeks before collision.

On December 9, Shoemaker–Levy 9's fragments made their first appearance in the morning sky, ready for the second and final act of their dance. Jim Scotti was the first to image them. Almost hidden in the glare of a bright star, the fragments had spread farther apart during their months of hiding. Five nights later, the University of Hawaii's David Tholen caught them through a large telescope atop Hawaii's Mauna Kea. Each fragment now had its own tail; or at least, the fragments were now separated by such a distance that their tails were clearly separated.

The comet's reappearance brought a collective sigh of relief.

Not that anybody wanted to mention it, but some observers shared a subliminal fear that the pieces would just fall apart and disappear. Comets closer to the Sun have sometimes done that, but except for two, J and M, the bodies approaching Jupiter were still holding together—a good sign that they were solid. The surrounding wings of dust had spread out so much, like gossamers, that they were far harder to see.

The most exciting part was that we now were following the fragments along longer arcs of their orbits. The longer the orbit arc is, the better understanding we have of the entire orbit. In mid-December, a new and improved orbit was ready. That afternoon I went over to Tucson's Planetary Science Institute, where the hot topic of a year-end party was the hardy band of comets out there by Jupiter. The prevailing wisdom, Galileo investigator Clark Chapman suggested, was that from the spacecraft's point of view the impacts would appear near the limb, or edge, of Jupiter. Maybe Galileo's eye would see the tops of the fireballs, Clark Chapman allowed, if they went high enough. (Fireball is used not in the sense of a bright meteor, as meteor scientists say, but as a fireball rising after an explosion.)

As the eggnog flowed, the discussion turned in a new and whimsical direction: If some people could couch S–L 9 in religious or national terms, why shouldn't we? An S–L 9 nation would have two special holidays each year, the March 25 discovery date and a seven-night-long festival every year to celebrate impact week. The S–L 9 national flag, we went on, would show a long string of pearls, perhaps with a quizzical Jupiter awaiting their arrival.

That cold afternoon, after the party ended, I drove home for a weekend rest. I walked through the door and decided to check my e-mail. Waiting for me was a fabulous new circular from Brian Marsden: The projected sites of all the impacts, he allowed, had moved much closer to the morning terminator. The collisions would still be on Jupiter's night side, but instead of waiting two hours to see the sites from Earth, we would see the results inside of 20 minutes. I could hardly contain my excitement. "Clark," I said to Chapman's answering machine, "there's great news. Galileo will see every one of the impacts!"

What wonderful news. The comets were creeping back toward Jupiter on much more favorable paths. If the tops of Jupiter's clouds could hold some evidence of the impacts for more than 20 minutes, then we might view some of the show directly from Earth.

BALTIMORE: FINAL PLANS ARE LAID

After working virtually round the clock to finish the draft of my new book *Skywatching*, I finished up 1993 in a state of exhaustion and then headed off to Palomar to spend a week observing with Gene and Carolyn. The weather was better than it had been a year earlier. This time viewing conditions were decent through most of our seven allotted nights. Near the end of the observing run, when Jupiter would be highest in the sky, we tried to record our comet. But its fragments had become too faint to detect, even on the sensitive emulsion of our film.

We wrapped things up early on our final night, because the next day we were flying to Baltimore for a major S–L 9 planning workshop. The idea was to arrange and coordinate all the world-wide observations of Shoemaker–Levy 9. To identify the key types of observations that needed to be made for each aspect of the collisions between S–L 9 and Jupiter, the meeting was organized into seven scientific working groups.[1] Although the meeting was intended to be a brainstorming rather than an academic session, for me the meeting was a good review of our state of knowledge of comets and of Jupiter. Before observations could be planned, theoreticians needed to predict what was likely to happen on Jupiter during impact week.

Most of the prognostications assumed that the impacting bodies would be about a kilometer in diameter, which seemed back in January to be a safe guess for at least some of the fragments. Mellosh and Scotti still felt that the largest pieces were not more than a kilometer across, while the Sekanina–Yeomans–Chodas team assigned that diameter to the smallest fragment. As debate continued on what to expect, the European Southern Ob-

servatory's Richard West—of 1976's great Comet West fame—
suggested that since the fragments seemed to be in a range of
sizes, everybody's predictions could turn out right.

THE ORBIT

According to diagrams prepared by Donald Yeomans and
Paul Chodas, the fragments of S–L 9 were traveling in a very
elongated orbit about the planet Jupiter. But their actual orbits
were even more extreme than the stylized sketch indicated: They
were traveling almost in a straight line away from the planet,
slowing down to a virtual stop, and then heading back. So at
apojove, the fragments lined up one after the other directly behind
Jupiter.

Although the earlier impacts would be just behind the morning limb, or edge, of the planet, the later impacts would be even
closer—so close, in fact, that if the explosive fireballs arched more
than 400 kilometers above the cloud tops, we'd be able to see
them. No one projected that we would see any plumes from the
earlier impacts, however.

Because we would watch a comet come apart, one of the most
significant things were hoping to learn from the crashes was how a
comet was put together. We knew that the comet broke up on July
7, 1992, at 1.3 Jupiter radii—a little more than half the diameter
of Jupiter—away from the planet. But how did the fracture pro-
ceed? When did it occur, at perijove or later, or earlier? Would the
whole comet crash into Jupiter, or would some of the particles in
the wings escape? According to Don Yeomans, not much of the
comet—some 10^{10} grams—would escape collision. This is the
equivalent of a body of dust about 25 meters across and might
weigh about 5000 metric tons, perhaps the size of a house.

Zdenek Sekanina of JPL expounded on his model of the size
of S–L 9's progenitor comet. The time of disruption, he maintained, was critical. If the comet fell apart about two hours after
closest approach to Jupiter, as he believes it did, then the original

progenitor comet must have been some 10 kilometers wide—virtually the size of Halley's comet. There is a difference between physical separation, when the pieces fall apart, and dynamic separation, which occurs when the pieces go their separate ways without any gravitational help from the others. It is likely, Sekanina conceded, that the comet tumbled apart precisely at perijove. However, for a little less than two hours after that event, the pieces were close enough to each other that their mutual gravity held them together. Finally each fragment began its own independent orbit about Jupiter.[2]

On that frigid January day in Baltimore, 160 scientists and reporters were trying to understand the magnitude of the upcoming event. Were the fragments solid bodies that would carve deep tunnels of fire into Jupiter, or were they something else? From Tucson's Planetary Science Institute, Stuart Weidenschilling had just proposed a new model in which comets were made up of house-sized pieces with room-sized spaces in between. Although Weidenschilling wasn't at the meeting, Michael Mumma of Goddard Space Flight Center argued Weidenschilling's point effectively. If this comet were indeed a rubble pile with large spaces, would it not just fall apart as it approached Jupiter? And if that were to happen, wouldn't Jupiter just get pelted by a storm of house-sized particles instead of 21 pieces the size of subdivisions? "What we're seeing," Mumma suggested, "is a swarm of particles that will continue to disrupt as it closes in on Jupiter."

This set off a big debate. The Jet Propulsion Lab's Paul Weissman argued that if the comets were that poorly held together, how could the earlier versions of S–L 9 that plowed into Callisto and Ganymede have left large craters? Surely those craters were the results of discrete, solid nuclei.

FRACTURE DYNAMICS: BREAKING UP AT MACH 100

Harvard's George Field, former director the University's Center for Astrophysics, echoed the feelings of everyone: This is a

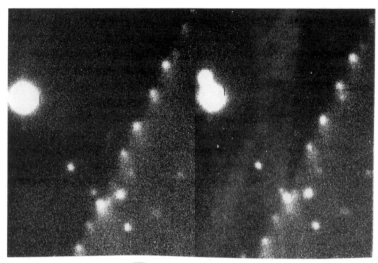

Earthquake! The University of Arizona's Steve Larson was imaging Shoemaker–Levy 9 on the morning of January 17, 1994, using the 90-inch Steward Observatory Reflector on Kitt Peak. Both pictures show the comet train, with the Q complex as the brightest section. The picture at right felt the earthquake in Northridge, California, shaking 500 miles away. The telescope shook so briefly that only the image of the bright star is affected.

very exciting discovery, he began. We stood to learn a lot about how cometary material is put together, he continued; but even more important, we would learn firsthand about what happens when an impact occurs. It couldn't be as simple, he went on, as studying the effect of an impact of a solid body on something like a wall. Because Jupiter's cloud deck is a very soft wall, the fragments would hurtle into it at 100 times the speed of sound. The collisional dynamics, he concluded, would be very complicated. The energy release, he proposed, would take place in two segments. Each fragment would drill an atmospheric hole in seconds, slowing down and tearing apart as its energy gets released into the atmosphere. The second release of energy would appear as the fireball that could take more than 100 seconds to dissipate. But all this would be moot, Field cautioned, if the fragment fell apart before it hit the atmosphere. The net amount of energy would be the same, but it would be released in a series of small pieces 10 meters wide or less, far too small to be detected from Earth or even the spacecraft Galileo.

JUPITER'S INTERIOR

What about the shock waves that would be produced when Jupiter got socked by all the energy of the impacting comet fragments? The most enticing of these waves would be the seismic waves. Just as we learned so much about the Earth's interior from the waves produced by earthquakes, scientists like the University of Arizona's Donald Hunten hoped to detect Jovian seismic waves from S–L 9. Mark Marley of the University of New Mexico agreed: We might see a seismic wave that would last some time after the impact. If we could record such a wave, we might be able to detect Jupiter's metallic hydrogen core. As seismic waves travel through a planet, they essentially bounce back whenever they encounter any change in density. Thus, if a wave were strong enough, it might even bounce off Jupiter's core, revealing for the first time precious knowledge of the inner parts of Jupiter.

SPECTROSCOPY

Spectroscopy—the practice of obtaining and studying the spectra of an object—would provide a tool to help us learn about the comet, both in the months preceding impact and at the moment of impact itself. A number of observers who specialized in detecting elements from the light waves they give off in their gaseous state had been trying to detect the telltale emissions of sublimating gases from the comet. Anita Cochran from the University of Texas' McDonald Observatory detected absolutely no emissions but, she explained, with the comet so far from the Sun,

"We don't expect to see much more than this." Even the Hubble did not detect any emissions. The next best hope was that these emissions might be noted during the impact explosions or right after them.

In retrospect, not detecting the emissions of elements should not have been a surprise. At Jupiter's distance from the Sun, it is so cold that only the most active of comets would have started sublimating the ices they contain. A comet with a poor supply of these materials might not be reacting to the warmth of the Sun to the degree that our instruments would detect a change.

THE COMAE

Could the dust clouds around each fragment become visible during the impacts as they reflected light from the explosions? Gene Shoemaker thought they would, and the University of Maryland's Mike A'Hearn suggested that the comae could brighten by a factor of as much as 50. If the comae didn't fade any more before the impacts, they could be observed reflecting light from the impacts if astronomers use a coronagraph, a tiny circular device placed in front of the telescope's focal plane to block out the intense light of the planet.

"It is tough," A'Hearn concluded, "but doable."

GOING IN

Hurtling into Jupiter at an average of 6.5 degrees beyond Jupiter's limb, the comets were predicted to strike at varying distances from that edge—the earlier ones about 10 degrees away, and the later ones less than 5 degrees. Kevin Zahnle from NASA's Ames Research Center proposed a pancake model, in which each comet would flatten out as the pressure around it increased faster in the front than at the sides. The incoming missiles would then grind to a sudden stop and explode. In Jay Mellosh's model, the comets would penetrate so deeply before exploding that the planet would essentially swallow them up with scarcely a trace. Caltech's Thomas Ahrens had the most optimistic suggestion. A fireball between 15 and 30 thousand degrees Kelvin would form. Now a highway to space, the entry tunnel would become the conduit for the fireball, which would blow out into the vacuum of space above Jupiter's clouds. Ahrens prognosticated that the expanding front of the fireball would be unbelievably hot—some 200,000 degrees Kelvin (the surface of the Sun is only 6000 K), but Field and others suggested it would be cooler.

In summary, Gene Shoemaker suggested that the following events would follow hard upon each other: A fragment would cascade into Jupiter (he called this the initial entry meteor). Then there would be a rising fireball, followed by strong heating of gas in Jupiter's stratosphere. Finally, particles would condense high in the stratosphere.

Reta Beebe of New Mexico State University's Astronomy Department continued the predicted sequence of events. (Reta Beebe has studied Jupiter there for more than 20 years. Hired by Clyde Tombaugh, the discoverer of Pluto, Beebe has run the department's planetary patrol program that Tombaugh set up. New Mexico State has, in fact, the largest archive of planetary photographs and CCD images in the world.)

After the first three minutes, Beebe predicted, we might see the material that was ejected in the fireball falling back in. This "fallback of the ejecta material" might last up to three hours. We could also look for seismic waves going through the planet over the next day, or atmospheric waves coming from the impact site over a period of several days.

JUPITER'S RING

Circling about Jupiter less than a planet's diameter away, Jupiter's ring might possibly act as a reliable reflecting surface for the lights from hits. The ring could function in two ways. First, since it could reflect the flashes and fireballs that were happening on Jupiter's far side, it would be a good idea to watch the ring at these critical times. Scientists already knew to check the moons for similar reflections, but moons would not always be in favorable viewing positions at the moment of each impact. Some portion of the ring would be always in just the right place.

The other possibility was even more enticing. Could additions of material from the comet actually cause the ring to expand in size, or could they make a second ring form? One model predicted that the ring would brighten steadily over the next few years. The brightening could be as little as a fraction of 1 percent of its total brightness, which was pretty low to begin with. But other calculations indicated that it could brighten by as much as 10 times over several years. Moreover, the ring might not recover its normal state for as long as a century. The material for the ring would come from the comet's ancillary parts—its comae, tails, and wings. These would not hit Jupiter, but they'd stay in orbit around it and eventually coalesce around the planet's equatorial plane.

JUPITER'S MAGNETOSPHERE

Extending as far as 50 times Jupiter's diameter around the planet, the magnetosphere could affect the comet's small particles of dust. Some scientists suggested that the dust in S–L 9's many comae could react with its ionized particles and brighten significantly. There was also a possibility that during the collisions, the magnetosphere could become so charged with cometary dust that it could briefly become visible. Scientist Melissa McGrath echoed the common belief that such a light-up of the magnetosphere was unlikely. She recommended follow-up observations of the magnetosphere through 1995. She also said that amateur telescopes, with their wide fields of view, would be particularly suited to studying the Jovian magnetosphere. That meant amateur astronomers, as well as professionals attached to universities and government agencies, could play another important role, in addition to the major work they did in providing precise positions of the comet in the months following its discovery. I felt a surge of excitement for the many amateur astronomers who, like me, spend long hours observing the sky.

CORONAGRAPHS AND CINE

The Comet Impact Network Experiment was the ambitious program of Steve Larson of the Lunar and Planetary Laboratory in Tucson. The idea of CINE was based on observing the comets approaching Jupiter, and their impacts, from sites at longitudes all around the world. As fluffy cometary dust enters Jupiter's magnetosphere, he thought, it will become charged and might brighten. We might actually see the comet's dust comae breaking up as they get closer to Jupiter.

Once big challenge was observing the comet fragments brushing up against the enormous brightness of Jupiter. The solution: Use a coronagraph. The Moon itself served as the earliest coronagraph. Because it blocked out the intense light of the Sun for a few precious moments of total eclipse, watchers could see the Sun's prominences and corona. The early artificial coronagraphs blocked out the Sun so that the faint prominences would become visible at any time. What Larson planned to do was to custom make coronagraphs for each of nine telescopes spread apart at various longitudes around the world, in northern and southern hemispheres. Just arranging for observing time on all these telescopes was a challenge. Each one was run in its own way, with different procedures and different deadlines for applying for observing time. By the Baltimore meeting, he was in the midst of this tedious process.[3]

THE SPACECRAFT FLOTILLA

Hubble Space Telescope

To everyone's delight, Harold Weaver of the Space Telescope Science Institute disclosed that the newly repaired Hubble Space

Telescope, with its vexsome astigmatism corrected weeks earlier, would allow some 100 of its orbits around Earth to observe Jupiter and the comet. Of those, 36 orbits—slightly more than a third—would be used to image the comet fragments as they approached Jupiter. If all went well, the Hubble would image some of the nuclei, notably Q, as close as a few hours before impact. If fireballs from some of the larger and later impacts were visible at all, the Hubble would be the ideal telescope to pick them up.

Galileo

At the moment of the first impact, the magnificent Galileo spacecraft would be 17 months away from its own encounter with Jupiter. Its position, so far away from Earth, was along a different line of sight to the planet, which would allow its cameras to record a slice of the giant planet's night side. Thus, the craft would have a direct view of the impact sites invisible to Earth. However, as Clark Chapman and Steve Edberg cautioned, the craft's tape recorder was limited. The timing of the impacts would have to be known precisely—to a few minutes if possible—in advance. And that would be the catch-22: By the time instructions must be sent to the spacecraft in early July, the impact times would still be known only to about a quarter of an hour.

The Galileo team was thus faced with a few other challenges. When the spacecraft was launched in 1989, no one knew that its main antenna was damaged. It was supposed to open in space like an umbrella. But the antenna jammed on opening, a failure possibly due to the wearing away of lubricant during repeated cross-country trips during its years of launch delays after the Challenger disaster. So Galileo was left with a low-power, wide-angle antenna that could send data back to Earth only at 10 bits per second, agonizingly sluggish compared to today's modems. Computer users know that a slow modem transmits at 2400 bits per second, and many home modems use 19,200 or even 28,800 bps. What could be done to maximize Galileo's return? Although only 20 frames could be stored on the tape recorder, the craft could store almost 64 images of Jupiter by multiple exposures, after re-aiming, on each frame. The strategy would be to take different types of pictures in a time window around several of the impacts. However, given Galileo's slow transmission speed, only a quarter of these crucial snapshots could be returned to Earth before the spacecraft's computer system was needed for its preapproach preparations with Jupiter at the end of January 1995.

The second, and very serious, challenge was how to control the spacecraft's camera when the times of the impacts would not be known more accurately than to a quarter of an hour. "You can't joystick a spacecraft," Edberg insisted. You need to plan its observations well in advance. Since the positions of the comet fragments relative to each other were known far better than the exact positions of the comet fragments at a given moment, one strategy would be to upload a command sequence that would start at a certain time, and then simply send along a revised start time for a whole block of observations. If the first impact could be timed accurately, then the specific times of the remaining hits could be calculated to an accuracy of about a minute.

The third vexing issue was whether light scattered by a crashing fragment could damage the instrument that tracks stars on the spacecraft. If that happened, the 1.5 billion dollar Galileo would lose its ability to position itself and point its cameras at the appropriate places in the sky, not just at impact time but for the rest of its entire mission.

Fourth, the high-resolution camera was not compatible with measurement schemes for the other remote sensing experiments. But by dividing itself into at least 21 pieces, the comet solved this problem nicely, allowing planners to divide the spacecraft's agenda among different investigations.

What kinds of observations was the Galileo spacecraft team planning? According to Steve Edberg, direct imaging of impacts was the most obvious type. The spacecraft could also take time-lapse, smeared images. At least two impacts would be recorded by leaving the shutter open as the platform on which the remote sensing instruments were mounted scanned slowly. The resulting image would show Jupiter as a series of wide, bright stripes, with the impact meteors and rising fireballs appearing and disappearing in Jupiter's dark night side next to the smeared daylit Jupiter as functions of time.

The photopolarimeter radiometer, a kind of light and heat meter, would be the first to return data on the impacts. The spacecraft's computer was instructed to collect and return its data on several of the impacts within hours of their occurrence. Sharing this relatively rapid data return was the plasma wave spectrometer, whose radio receivers were listening for changes in Jupiter's radio noise output. The complement of remote sensing instruments included an ultraviolet spectrometer and a near infrared mapping spectrometer. These instruments, along with the photopolarimeter, were allocated some tape on the spacecraft's tape recorder to make complementary measurements across a wide range of the electromagnetic spectrum.

Lastly, as the spacecraft approached Jupiter over the following 17 months, it could search for and measure any increase of dust ejected by Jupiter as a result of the comet collisions. Whatever the timing of the fragment impacts, Galileo would play a crucial role in maximizing what science would learn from the event. What luck that the mechanical namesake of the revolutionary Renaissance stargazer was out there and headed Jupiter's way.

Voyager

A collective groan rose from the audience at the Baltimore meeting when we heard that Voyager 2's camera had been so permanently shut off that there was no hope of turning it back on. When the turn-camera-off decision was reached a few years earlier, the software was erased: The camera's bank account was permanently closed.

Although Voyager's hard luck story was disappointing, the spacecraft team hadn't quite reached the end of its trail. The craft's radio experiments were still working and might add important data, even though the craft was 40 astronomical units from the

Earth, and Jupiter's image would take up only two picture elements (pixels) on its high-resolution camera.

Ulysses

On its way to an orbit around the poles of the Sun, the spacecraft Ulysses carried a powerful radio receiver capable of detecting possible signals as each fireball rose above Jupiter's clouds. The good news from Ulysses was that it could accommodate a program of Jupiter observations without interrupting its mission of studying the Sun and its vicinity from a look above its poles.

Clementine

Finally, the Clementine spacecraft, about to start mapping the Moon, would be available for imaging and other measurements during encounter week. Lucky for us, Gene Shoemaker was the spacecraft's science team leader. Since that the craft would have finished its mapping mission by then and would be on its way to a flyby of the asteroid Geographos, it could turn its high-resolution camera toward Jupiter and just watch. "Who knows?" he said, "if the dust coma scatters the light from a flash, maybe Clementine will see it." Sadly, in June the spacecraft was undergoing propulsion tests when a computer failed. As engineers frantically tried to reboot the computer, the spacecraft emptied its supply of maneuvering fuel and was left spinning uselessly in space.

SENSE OF THE MEETING

Just as the October meeting in Colorado energized everyone, this early January gathering turned the heat up another notch. Baltimore seemed to awaken the media to the importance of the coming event, and journalists from all over the world began to cover it. The comet was now on the final leg of its last orbit.

In the past few months, I had felt that as if the story shifted from the discovery to the coming impacts. Still, as Baltimore ended, Lucy Ann MacFadden, who with Mike A'Hearn had organized the meeting, approached us and said, "Thank you for finding this comet! Your discovery has really generated a lot of excitement."

The meeting's ultimate success was generating the feeling that nobody wanted another meeting before impact week. "Should we have a follow-up in May?" No, came the answer, virtually unanimously. We'll be too busy preparing our observations by then.

SHOEMAKER–LEVY 9 MEETS THE PRESS

On February 25, Miles O'Brien, a science journalist from CNN, sent me an e-mail message asking for details about where Gene and Carolyn and I would be during the impacts. "We are planning major coverage," he wrote, "and are hoping that the three of you will be a part of it." Some months ago the three of us had decided to join forces during impact week. As head of the scientific team for the Clementine mission, Gene was planning to spend those days at a place called the Bat Cave, Clementine's control center in Alexandria, Virginia. The Planetary Society had invited me to its shindig in Washington, but I thought I'd rather be with a telescope and some clear sky out on a mountaintop in California. "Let's be together that week," Gene argued. "Crash week is a time for the three of us to stay close." So I accepted the Planetary Society's invitation.

A short time later, writer Jim Reston called to make appointments for interviews for a major article he was writing for *Time*. And by early April, when *Time*'s science editor Michael Lemonick called with further questions, we knew that the news magazine was planning a major story. The media had figured out that this might be the biggest astronomical news in some time, caught wind of the story, and was intrigued. We hoped that the show would not be disappointing.

Slings and Arrows of Outrageous Fortune

As Shoemaker–Levy 9 cruised smoothly toward Jupiter, people on Earth were wondering how often the Earth gets similarly spectacular—and dangerous—celestial treatment. The answer, of course, depends on how big the original comet was before it fell apart in 1992. The Earth has been hit repeatedly in the past and will be hit again. We are a dartboard in space, with enough slings and arrows aimed randomly at us that we do have the outrageous fortune to be walloped from time to time.

Gene Shoemaker, who makes the study of impacts his life's work, estimates that our planet gets hit by a 10-kilometer-diameter (about 7 miles) comet or asteroid every hundred million years on average. More modest masses, a kilometer across or larger, hit our world more often, say once in a hundred thousand years. These odds sound pretty small, until we express them this way: There is about a one in a thousand chance that such a space traveler will hit the Earth within the next century. The threat hangs over us like a sword of Damocles: For annoying his master Dionysus, the mythical slave Damocles had to sit at a table with a sword hanging by a thread over him, ready to fall at any time.[1]

WHAT IF A COMET OR ASTEROID WERE HEADED HERE?

If someone were to discover an asteroid that seemed to he headed toward us, astronomers would observe it with telescopes around the world to determine its precise path with great accuracy. The Minor Planet Center would act as a central clearinghouse, handling the collection of observations and the initial publication of orbits. Other institutions, particularly the Jet Propulsion Laboratory, would calculate orbits as well. After the maelstrom of observations was sorted out, in all probability the new orbits would have the intruder grazing past us as a near miss, and the scare would be over. (Imagine what a field day the supermarket tabloids would have!)

But what if after all the best measurements, the object still seemed headed directly toward an eventual collision with Earth, perhaps 30 years in the future? Then it would be prudent to send a spacecraft to confront the intruder as quickly as possible. The craft would photograph the object, so that we would understand its shape and composition. We'd need to know what we were dealing with. Maybe even the slightest nudge from a nuclear weapon would tear the object apart, causing us to lose any semblance of control over it. Finally, the craft would then plant a transponder, a signal generator, on this interplanetary hunk so that we could track its future course with great accuracy.

Finally, we could explode a nuclear warhead at a distance of about a kilometer from the intruder. The idea would not be to break it apart, but to blow off a thin layer near the object's surface to nudge it away from its threatened encounter.

Our ability to defend ourselves in this way would depend on someone's finding the intruding object at least a decade before its destined rendezvous with Earth, so that we would have sufficient time to mount a deflection campaign. Current thinking involves making all this fuss only for an intruder a kilometer in diameter or more. If the comet or asteroid is much smaller than that, it would be very hard to find it in time. Also, while the damage it would cause would be severe, it would be limited to the local area around which it hit. But if the intruder were about a kilometer across or more, huge amounts of dust thrust into the upper atmosphere by the impact would engender a more sinister scenario. Over a few hours, the dust would spread itself over the globe, cutting off sunlight and slowing down photosynthesis in a process similar to the familiar model of nuclear winter. Death would cycle up the food chain, from plants to predators and omnivores like us.

If the comet or asteroid were much larger, say 10 kilometers in diameter, the devastation would be so bad that mass extinctions of most species, including possibly our own, would result. During our many interviews with the press during 1994, the media lobbed this question at us so many times I lost count. Obviously the topic caused a lot of anxiety. During a taping in May for the BBC children's program *Blue Peter*, Carolyn Shoemaker was asked what would happen if a 10-kilometer-diameter comet smashed into Earth. Looking directly into the camera, she described the sequence of events and ended with a terse, "We would die." The producer frowned at her answer, which she thought inappropriate for a children's show. "Carolyn," she said, "you need to get away from doom and destruction." Carolyn thought about this for a while, trying to find a less direct way of stating the obvious. "The impact would put a lot of material into the air," she began, "and it would be very dark." Finally, she concluded, "We would be very uncomfortable."

This became a standing joke for that oft-asked question, "What if S–L 9 were hitting the Earth?" We would indeed be very uncomfortable. Say that the size of the fragments averaged 1 kilometer and that the first fragment hit off the coast of disaster-prone southern California. The immediate destruction would be immense. Then dust from the gouged-out crater would rise right out of the atmosphere and land again at sites all over the Earth. High in the stratosphere, large amounts of dust would cut off sunlight and set off a slowdown of photosynthesis. If the impact occurred at the start of spring in the northern hemisphere, which produced most of Earth's crops, an entire growing season could be lost worldwide.

All this damage would result from a single 1-kilometer chunk. But six hours later a second would hit, at the same latitude but at a much different longitude, maybe over the Atlantic. Another would hit over Tucson, yet another over a city with similar latitude, like Tel Aviv, and a final one over Los Angeles again, for good measure. By the end of a frightening week the Earth would have been struck by more than 20 comets. The amount of dust in the atmosphere would now be so great that the Sun's light would be almost completely shut out for months. The effect, like shrapnel, would actually be worse than if the whole comet had hit us as a single large object.

Fortunately, the chances of even one of an S–L 9 type fragment hitting the Earth are extremely slim. The odds of any individual's dying from a comet or asteroid hit are estimated to be about one in 20,000. This is about the same as the chance an average person living in the United States has of dying in the crash of a commercial jetliner. Compare this with perishing in a car crash (your chance is a high 1 in 100) or being struck by lightning (not much: 1 in 4 million).[2] Of course, these figures don't take into account the variation of risk due to behavior. During a bad thunderstorm in Phoenix, a reckless man walked outdoors to light a cigarette and was promptly executed by a lightning bolt. Had he not been struck, his chances of dying from lung cancer would have been as high as 1 in 10.

These statistics have another caveat to take into account, and that is risk versus consequence. Although the risk of dying in a comet crash is low, the consequences of a crash are very high, since a single impact could kill billions of people or, in the extreme case, everyone.

WHO LOOKS FOR COMETS AND ASTEROIDS?

In the fall of 1952, Gene and Carolyn Shoemaker visited the large crater on Coon Butte, outside Winslow, Arizona. To Gene, the crater appeared to have formed by some sudden convulsion, but exactly what was unclear. Back in 1903, Daniel Barringer suspected that the crater was of impact origin, so he put up half a million dollars to buy the land around the crater and start a search for the big iron meteorite. The large mass was never found: It was never there, since most of it had vaporized on impact. Although people from Barringer onward suspected it to be an impact crater, it was Gene Shoemaker who demonstrated its impact origin beyond any reasonable doubt. In addition to evidence provided by the structure of the crater and its surroundings, including layers of rock completely overturned, Gene Shoemaker and Beth Madsen, his colleague, found along its walls abundant samples of a mineral called coesite. Named after physical chemist Loring Coes, this mineral forms only under the levels of very high temperature and pressure that exist at the moment of an impact, conditions too extreme to occur during volcanic explosions or any similar cataclysms.

By the early 1960s, Shoemaker had set his sights on the Moon. Peppered with the circular depressions resulting from ancient impacts, our satellite carried a record of the kinds of events that Earth had eroded away. If our nearby Moon was a target, Earth must have been one also. During the middle and late 1960s the Ranger and Surveyor probes studied the Moon's cratered surface, and the Apollo astronauts prepared to study the Moon's geology under Gene's expert tutelage. During the 1970s his study of impacts went in two directions. One was off in space, as a member of the team working with the twin Voyager spacecraft to study the craters on the moons of the outer planets. The other was at Palomar Mountain Observatory, where he sought to find the comets and asteroids that make the craters.

There are 2000 Earth-crossing asteroids larger than a kilometer, Gene Shoemaker estimates. His search for asteroids began in

1972, when he and Eleanor Helin began using Palomar's vintage 18-inch telescope to find asteroids that can ultimately collide with Earth. At the time, only about a dozen such objects had been identified. Ten years later, the two scientists went their separate ways, mounting different search programs. Meanwhile, the University of Arizona's Tom Gehrels began an independent search program combining technology old and new—an old 36-inch-diameter telescope fitted with a brand-new CCD system. Based at Kitt Peak National Observatory near Tucson, this Spacewatch program has been the work base for Jim Scotti, along with Tom Gehrels, Robert Jedicke, and, until recently, David Rabinowitz. So far, they have found objects as small as a few meters in diameter whizzing by the Earth. In Australia, Rob McNaught has found many near-Earth objects on plates taken at a large Schmidt camera.

These searches have found most of the 200 known asteroids that are now known to cross Earth's orbit. (We have come quite a way since 1972, when only about a dozen such asteroids were known.) The Shoemaker and Helin teams watch the sky for a week each month, and the Spacewatch program uses its telescope for three weeks each month, centered around new moon. However, if Gene's estimate of 2000 is correct, a lot remain for us to find and track.

SPACEGUARD: TAKING INVENTORY
OF THE NEIGHBORHOOD

By the end of the 1980s, astronomers became aware that the Earth and its inhabitants were in some danger from asteroid or comet hits. But we had only a statistical idea of the magnitude of that danger. Our chance might be very slim for a kilometer-sized comet or asteroid to hit us this year, but if something sizeable out there is eight months away from a collision with Earth, then it's no longer a statistical problem. What was needed, many scientists felt, was an organized, global effort to find the remaining 95 percent of the objects bigger than a kilometer and determine whether any of them was on a collision course with Earth in the near future.

A few years ago, NASA formed a committee chaired by David Morrison of NASA's Ames Research Center to find a way to solve this problem. They came up with a plan to mount a series of telescopes around the world. Together these would scan the sky so thoroughly that 95 percent of the potential kilometer-or-larger impactors would be charted within 25 years. In addition, the program could find many of the 300,000 asteroids crossing our orbit and larger than $\frac{1}{10}$ kilometer that could cause severe local damage if they hit.

Locating potentially impacting comets posed a much more challenging problem. Since comets orbit the Sun in long, looping orbits, they tend to be visible for only a few months to a year or more out of their revolutions, which in turn could last from a half dozen years to hundreds of thousands or millions of years. As any amateur comet hunter knows, a comet search would have to be ongoing, always finding more comets but never completing the task.

This theoretical problem suddenly seemed more urgent in the fall of 1992, when Comet Swift–Tuttle returned to the inner solar system for the first time since 1862. Famous because it is the parent of the Perseid meteors that put on a splashy display every August, this comet's return was an exciting event. It represented a triumph for Brian Marsden, who in 1973 had predicted its return to within about two weeks.[3] Each night observers could see it brighten as it swung through the northern sky.

In October 1992, Brian Marsden issued an IAU circular with a new and more accurate orbit for Comet Swift–Tuttle. He had long agreed with the earlier researchers Oppolzer and Schiaparelli that the comet could make a very close approach to the Earth if it rounded the Sun in late July of any year. That makes sense. Since the comet's trail of dust encounters Earth every year, maybe the comet itself could hit the Earth someday if it, too, happened to intersect our orbit when the Earth was there.[4] The new orbit, based on observations made as far back as 1737, suggested that at its next return, the comet would round the Sun that July 11.[5] But what if the comet were late, returning to perihelion during a fateful three-minute period on July 26, 2126? Then, Marsden suggested, the 15-kilometer-diameter comet could crash into Earth on August 14.[6]

When this circular came out, the press responded with massive coverage of the impact hazard; *Newsweek* even ran a cover story on Comet Swift–Tuttle.[7] But a few months later, after the comet's orbit was traced backward, Brian announced that the comets of 69 B.C. and 188 A.D. were actually ancient passages of the famous Swift–Tuttle. Now the orbit was well enough known that there was absolutely no chance of a collision in 2126. However, Brian warned the world against letting down its guard: "Even if Swift–Tuttle doesn't get us next time," he wrote, "it is the largest object we know of that has the potential to strike the Earth over an interval of 10,000 years or more."[8]

A few months after Swift–Tuttle was on its way to the solar system's outer reaches, more than 100 scientists met in Tucson at a conference called "Hazards due to Comets and Asteroids." The conference dealt not just with the threat itself but the related issue of deflecting potential intruders with nuclear weapons. Edward Teller, inventor of the hydrogen bomb and advocate of the strategic defense initiative, popularly dubbed Star Wars, suggested using nuclear weapons to move asteroids about to gain an understanding of how the process would work. If an intruder were actually on its way, we would have an idea of what effect our attempts at deflection would have. Since we didn't even know if Damocles existed, critics howled that the idea of deflecting such a comet or an asteroid was merely a make-work project for an aerospace industry badly hit by the end of the Cold War. Editors and cartoonists raged against the idea of deflecting asteroids and comets. An editorial in *Nature* chided: "Despite all the hopes pinned on it," they wrote, "Swift–Tuttle is resolutely refusing to be the agent of wild destruction. Too bad!"[9] This "giggle factor"—as Gene Shoemaker called the failure, at the political level, to take the impact risk seriously—threatened to derail the entire program, including the Spaceguard program to find these objects with their telescopes. Meantime, the weather added to the gloomy mood of the scientists as record-breaking rains flooded out parts of Tucson.

At that meeting, the Shoemakers and I met Philippe Bendjoya, who was studying a problem related to the issues being discussed. In the long history of the solar system, asteroids have collided with other asteroids, breaking apart into families of smaller asteroids. A billion years later, how do we trace back the orbits of asteroids to a common parent before it broke apart? This is a daunting task, and a number of scientists, including our *Niçois* friend Bendjoya, are investigating it. He was interested in learning how we go about our search for comets and asteroids. In fact, he asked to join us on our observing run at Palomar set to begin in less than two months' time, in March of 1993. We could not have imagined then that the observing run that we were discussing with Bendjoya would have a bearing on collisions of comets with planets. We could not have imagined that at our March run, we would find a comet that would thrust impacts into the consciousness of humanity.

A Disabled Ship

Like a disabled ship heading helplessly for a barrier reef, Shoemaker–Levy 9 hurtled toward Jupiter. Crash week was approaching fast, and thanks to a stunning series of comet portraits sent down from the Hubble Space Telescope early in 1994, public interest was booming. The first of the 1994 images came at the end of January.

Using the image-processing software at the Space Telescope Science Institute in Baltimore, Harold Weaver welded three of the images to form a mosaic that included all the comet fragments except the last two, V and W. His composite was a work of art, showing the comet fragments with their tails dancing in line like the Rockettes. It was especially satisfying to compare it to the July 1993 image. The images were clearer and showed what the corrective optics had done for Hubble.

A close look at the two images, however, showed that the change in the actual appearance of the comet was profound. The difference was due to more than just improved telescope optics: The comet itself was undergoing dramatic change. In the July 1993 likeness, fragment Q appeared to be splitting apart. Six months later, Q was clearly two components, which we called Q1 and Q2. Zdenek Sekanina of the Jet Propulsion Lab later suggested that the reason that Q was so bright was not its size but its dynamism; it had torn in two just a few weeks after the comet's discovery. This was not the first breakup of Q since the original disruption, when the comet passed Jupiter in July 1992, Sekanina added. P broke off earlier, and then subdivided into P1 and P2. Meanwhile, B separated from C, and F dissociated from G. The daughter fragments that split from the original comet are all on a line. The granddaughter ones, on the other hand, lie noticeably off that line.[1]

HST's March image brought further surprises. In just two months, more disruption was taking place. Fragment P1 split in two, and nothing visible was left of P2 but a faint puff of what was probably cometary dust. That image raised an important question: Would there be *anything* left of the disabled ship by July, other than some loose wood? Would the comet disintegrate into a large group of tiny particles? Some scientists thought that Shoemaker–Levy 9 would fizzle and that there would be no big explosions on Jupiter.

"There's no evidence for nuclei of any size," said the University of Maryland's Michael A'Hearn when asked another important question: how big comet S–L 9 really is. Because no telescope could penetrate the thick layer of dust that surrounded each nucleus, we were unable to pin down the size of each one. Misinterpreting A'Hearn's statement, a surge of press reports announced that there were no nuclei at all and that Earth-bound observers would see nothing at impact week. No. All that A'Hearn meant was that the Hubble's imaging had not penetrated the dust comae and did not see any nuclei. Some big cometary fragments could still lurk underneath a layer of dust. The HST data by themselves, Hal Weaver concluded, could not be used to state that the nuclei were between 1 and 4 kilometers across. It was possible, he went on, that all of the nuclei could be considerably smaller than a kilometer.[2] The fact that HST didn't see the nuclei meant that the size debate would go on right up until impact time and that the

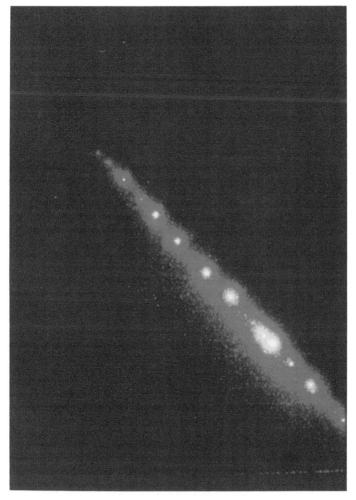

The first exposure of Comet Shoemaker-Levy 9 taken from space. In early July 1993, the Hubble Space Telescope took this image of the shattered comet.

predictions for what might happen would remain agonizingly vague.

Both HST and images from ground-based telescopes showed that fragments J and M had completely disappeared from view. What had happened to them? Had they somehow avoided Armageddon? The most likely theory: They were swarms of small fragments so loosely held together that they fell apart. Did that mean that all the fragments would fall to nothing before crash week?

Large or small, the comet fragments were closing in on Jupi-

Like a group of kicking Rockettes, the fragments of Shoemaker–Levy 9 perform for this Hubble Space Telescope image, taken in early January 1994. Courtesy H. Weaver, Space Telescope Science Institute.

Comet Shoemaker–Levy 9, photographed using a red filter by the Hubble Space Telescope on May 17, 1994, just two months before the first impact. When the January 1994 images were obtained, only three could cover the entire train. But now the comet train stretched over 710,000 miles, or more than a million kilometers, and six exposures were required to capture all the fragments. Courtesy of H. A. Weaver and T. E. Smith, Space Telescope Science Institute.

ter, and by the middle of April a line of reporters for print, radio, and video interviews were closing in on the doorstep of our 18-inch telescope. Soon this became an almost daily onslaught. Although not used to such attention, Palomar Observatory quickly adjusted to the increasing flow of reporters and their requests. Carolyn and I found the pressure a lot of fun, but it was stressful. Since Gene was off near Washington working with the Clementine mission, then on its way to map the Moon, he escaped the trail of media types who dropped in on our March and April observing runs.

In April, the NBC nightly news broadcast Mordecai-Mark Mac-Low's beautiful computer simulations showing a 1-kilometer-diameter comet exploding on Jupiter. On Tuesday morning, May 17, S–L 9 was on the cover of *Time* in a vivid painting by artist Julian Bacon. Written by astronomical historian James Reston, son of the famous *New York Times* columnist, the *Time* story emphasized the people involved in the discovery and in the complicated joining of observation and mathematics that led to the calculation of the comet's fatal path. The following day NASA put on a press conference. Held against a large backdrop showing the Hubble Space Telescope's January image of the comet, the conference was chaired by Gene Shoemaker and included the major players on the HST science team. *Time*'s story drew attention to the press conference, which was widely covered. For more than a week afterward, wherever I went I saw images of Gene and other members of the HST Jupiter team flashing across TV screens in hotels and in airports.

During this exhilarating and chaotic time, our personal lives were getting more and more turbulent. Reporters descended on our homes. In one frenzied week CNN visited my home at almost the same time as the BBC visited the Shoemakers. In April I began a lecture tour that would climb to almost a hundred talks across North America and that coincided with my two new books, *The Quest for Comets* and *Skywatching*. At the end of June I started book signings for both these books in several major cities. As impact week approached, the crowds at these lectures and sign-

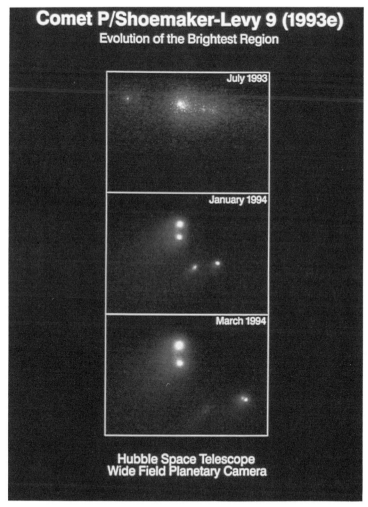

The fragments continue to fall apart, as shown in this sequence of Hubble Space Telescope Images. Courtesy H. Weaver, Space Telescope Science Institute.

ings increased dramatically. As my privacy dwindled, I was beginning to understand why some people do not like public attention.

To the alarm of some scientists, public expectations were rising way out of proportion to the comet's predicted performance. "The Big Fizzle is coming!" trumpeted Paul Weissman in *Nature*, partly in an attempt to cool off the excitement.[3] In a lively piece of writing, Weissman summarized a rubble pile model by Asphaug and Benz in which S–L 9, like most comets, was composed of material so loosely held together that the slightest tug would pull it apart.[4] If this were true—and HST's images showed that the comet was continuing to fragment—by the time it reached Jupiter there might be no large objects left to impact. A storm of meteors, dramatic on Jupiter but not visible from Earth, might be the extent of the show. "The impacts will be a cosmic fizzle," Weissman predicted. "The Shoemaker–Levy 9 explosions may be closer to about 30 megatons each, but still far less than the 100,000 megaton explosions that some have predicted."[5]

Brian Marsden was almost certain that S–L 9 was a string of small objects. "It is difficult to argue statistically," he wrote in March. Confirming his feeling in June, he stated, "But I think there are many small short-period comets in the Jupiter–Neptune region and that a fair number are temporarily orbiting Jupiter at any time. The bulk of these objects are less than a kilometer across. 1993e has done a strange thing by passing close enough to Jupiter to break up on one perijove and actually to collide with Jupiter on the next pass. Odds would tend to favor a small object doing this."[6] Marsden buttressed it later with the prediction that kilometer-sized comets actually collide with Jupiter every 15 years. "We never see Jupiter affected by these collisions," he added.

O. J. SIMPSON vs. SHOEMAKER–LEVY 9

On, Friday, June 17, accused of murdering his ex-wife Nicole Simpson and her friend Ronald Goldman, O. J. Simpson and a former teammate drove a white Ford Bronco sedately along L.A.

freeways trailed by dozens of patrol cars. The parade was broadcast live to millions of people. In the month between that drive and
the first comet impact, media interest in the plight of S–L 9 rose
and fell depending on what was happening with the Simpson
case. One interview I had with a radio station in New York was
almost canceled when the judge made a ruling in a minor issue
relating to Simpson's preliminary hearing. The station considered
discussing instead the decision: how Simpson and his legal team
were reacting to the decision, how the prosecution was reacting
to the decision, and how everybody else felt about the decision.
Happily the station decided merely to report the decision and then
get on with Shoemaker–Levy 9.

   During those cluttered days, the Shoemakers and I, and many
astronomers around the world, were trying to use the comet
impact to encourage public awareness of the night sky. I felt that
this was the best opportunity in a long time to accomplish this
lofty goal. Maybe S–L 9's death throes would be just the ticket to
introduce the night sky as a user-friendly place. But in the first
10 days in July media interest sputtered according to the ebb and
flow of new developments in the Simpson case.

LAST ORBIT, FIRST METEOR

   Around the first of July, the leading edge of the comet train
started to encounter the gigantic magnetic field surrounding Jupiter. A few days later the forward dust wing began a celestial
fireworks show that, could we have traveled to Jupiter and
viewed it there, would have dwarfed any meteor shower ever
seen on Earth. For an hour or so every night, a hypothetical Jovian
observer could have witnessed a startling number—perhaps
millions—of meteors. Behind them in the sky would appear a
train of comets that brightened with each passing night.

   From the Smithsonian Astrophysical Observatory in Cambridge, Massachusetts, Brian Marsden was revising his orbit calculations for S–L 9's fragments. Across the continent at Caltech's
Jet Propulsion Lab in Pasadena, California, Don Yeomans and

Jupiter and Shoemaker–Levy 9. This computer montage was prepared by D. Seale of the Jet Propulsion Laboratory, using two Hubble Space Telescope images taken at different times of the planet and the approaching comet fragments. Courtesy D. Seale, D. Yeomans, and P. Chodas.

Paul Chodas were also busy revising their orbit computations. With new observations coming in from many observers, especially Jim Scotti, by early July the figures were changing slightly. Each new calculation produced a different set of impact times. As the orbit of each nucleus was being calculated independently of the others, some times would get earlier and others would be later, sometimes by as much as 20 minutes.

Observers moving to their remote sites all over the world depended on these predictions, and they anxiously looked at each new set to see if their chosen observing sites, which they had necessarily picked months earlier, were in better or worse position for a particular impact.

With each week the predictions got more precise. On July 5 Yeomans and Chodas published the set that the Galileo spacecraft was to use in its long-planned observing program.

Ten days before the first impact, a team of scientists and engineers for the Galileo spacecraft now on its way to Jupiter sent the first of two sets of commands to the space traveler. For the first time, Galileo was told that Jupiter needed to be watched carefully starting on July 16. The impact flashes were expected to take up no more than a few picture elements or pixels (the planet itself would take up 60 pixels). If the spacecraft took its pictures at the right time, they should be bright enough to show up nicely. Mission engineers uploaded the second set of comet commands to Galileo on July 8. Poised to turn toward Jupiter, the spacecraft was arming itself as a full member of the flotilla now on the comet's trail.

HUBBLE HICCUPS

Two weeks before the first impact and a hundred miles above the Earth, the Hubble Space Telescope, working perfectly since its repair half a year earlier, sensed that something was wrong and, as it had done around the time it was observing S–L 9 in July 1993, shut itself down and assumed its fetal position. The possibility that the Hubble would not be operating during impact week, being reduced to a late-night talk show joke, stung the engineering and science teams that had worked so hard for the spacecraft's success. Two days later the telescope's cover suddenly closed. As days passed without a solution, patience gave way to real concern. Not having HST operational during crash week was unthinkable.

The crisis ended almost as quickly as it had started. Shortly after engineers traced the problem to some faulty boards and bad software, the telescope was awake again and doing science. To everyone's excitement and relief, the Hubble Space Telescope was alive and doing the observations it was designed for.

OFF TO ISRAEL

After months of preparation, Jim Scotti was now on his way to Israel's Wise Observatory to observe the crash with their reflector. He was a member of Steve Larson's Comet Impact Network Experiment (CINE) Team, which observes the comets and Jupiter from sites around the world. Having flown from Tucson to Los Angeles, Jim struggled with a problem he found far more difficult than tracking a comet out by Jupiter. Lost at the Los Angeles airport and loaded down with luggage and a heavy observing instrument, he was navigating his way from one airline terminal to a distant one at the other side of the sprawling airport. Finally, he tried to pilot his observing apparatus through the airport's security screens. Never having seen such complicated equipment, the security guards were curious about the instrumentation they had just X-rayed. Jim described the purpose of his coronagraph, which would be attached to the telescope and blot out the image of Jupiter so that he could follow the comets on their way in.

SATURDAY, JULY 9

For me, each day in early July was crammed with activity as I tried to promote public interest in a crash on Jupiter and my two books on Earth. On July 9, I arrived in Philadelphia for two lectures and three book signings. George Cruys, The Nature Company's director of public relations, accompanied me during this trip and arranged my complex interview schedule. Before O. J. Simpson's preliminary hearing ended, Cruys had his hands full trying to persuade the media that Jupiter's story was as least as newsworthy as O. J. Simpson's case. After July 9, when the preliminary hearing ended, media interest suddenly gathered so much momentum that we were having trouble fitting everyone in. During the first three weeks of that momentous month I would be interviewed about 125 times.

TUESDAY, JULY 12

George Cruys and I had a full schedule in Atlanta that included a book signing and an afternoon at CNN, where science editor Miles O'Brien was interviewing me as part of that network's intensive comet coverage. Taking advantage of my presence at CNN's towering headquarters, two programs from the Canadian Broadcasting Corporation put me on their shows. When CNN's New York bureau realized that I was in the building, they set up a live interview for their 6 p.m. national news.

While waiting for the news to begin, Miles informed me of a rumor that the U.S. would invade Haiti in just two days. I looked at him in despair. "On Thursday? Two days before the first impact? Can't they wait a week?" I did not want to minimize the plight of the Haitians, but the timing would have been devastating for public awareness of S–L 9. I knew that Miles, who had put a lot of effort into the comet story, was nonplussed by this rumor also. "What," I asked, "would that do to CNN's coverage?"

"David," Miles said ruefully, "Jupiter could fall apart into 21 pieces and crash into Mars, and we wouldn't cover it." Neither would anyone else, I knew.

Meantime, back at Palomar, Gene and Carolyn were observing with Henry Holt. Halfway around the world, Jim Scotti was having difficulty persuading his new telescope to accept the coronagraph he had brought, but after a little tailoring the instrument was mated to the telescope and tested on Jupiter.

WEDNESDAY, JULY 13

In the midst of a grueling schedule of interviews and a book signing, more bad news arrived. An article from the Associated Press accused The Nature Company of deliberately misleading its customers about the comet crash. Since I had proofread every sentence of their publicity, this attack infuriated me. The story quoted two astronomers who claimed that they saw a sign in one of their stores that offered unparalleled views of the impact sites with even their smallest telescope, an instrument with a lens less than 2 inches in diameter. I knew that the signs did not say any such thing. "Shaken loose by a passing star," the largest sign trumpeted, "shattered by the forces of gravity, Comet Shoemaker–Levy 9 is heading straight for Jupiter. The countdown has begun." Using S–L 9's plight to get people excited about the sky could be the lasting legacy of these impacts. "This is not just the professional's comet," Cruys maintained, "this is everybody's comet." We hoped that the most important effect would be felt not on Jupiter but right here on Earth.

Half a world away, Robert Marcialis had just arrived at the observatory at Mount John in New Zealand. Like Jim Scotti, he was a member of Larson's CINE team. "God, it's cold!" he wrote about the mountaintop midwinter night. There he met a married team of long-time comet observers, Alan Gilmore and Pam Kilmartin. Their first night was clear, and they looked at the Eta Carinae nebula, the Jewel Box cluster in the Southern Cross, and some of the other southern hemisphere beauties that are the envy of northern hemisphere dwellers.

THURSDAY, JULY 14

Bastille Day, July 14, was the morning of the rumored Haiti invasion. I woke up in Atlanta to a call from a radio station I used to listen to when I was growing up in Montreal. The fact that they were calling me at all was great news. Maybe there wasn't an invasion after all. After the call I opened my hotel room door to see a copy of *U.S.A. Today* with my picture on the front page. Haiti was not the cover story; the comet was! (In fact, the military strike on Haiti never did occur.) George and I flew to Birmingham, where a Nature Company book signing attracted hundreds of people. But by afternoon, the same reporters who had interviewed me rushed back to cover a mounting storm of publicity against The Nature Company. They poked cameras in front of customers, demanding to know what promises store clerks had given them about the performance of their telescopes.

It was an eerie scene. The resolving power of a small telescope—its ability to bring out small details on Jupiter—was suddenly a matter of major public interest and was front-page news. It was nationally covered by print, radio, and television media. Now a big news story, the crash of S–L 9, along with everything conceivably related to it, was taking on a life of its own. Unfortunately this particular tangent, coming less than 48 hours before the first impact, was a real trial for me. Exhausted and depressed, George and I ended the day with several hours of plane delays in Birmingham and Atlanta. We finally arrived at our hotel in New York City at 3 a.m.

FRIDAY, JULY 15

Three hours of sleep were all we had in New York. George greeted me at 6:30 a.m. and quipped, "We're going to have a better day today!" On the West Coast, Gene and Carolyn finished their last night of observing at Palomar and returned to a group of reporters from NBC waiting for them in their living quarters. We met—electronically—as guests on NBC's *Today* show, and from sheer adrenalin talked excitedly about the impacts and their meaning. Gene and Carolyn were scheduled to get a little sleep and then fly east for impact week. Immediately after that live appearance, George and I flew to Washington to be met by a lengthy story in *The Washington Post*. Written by Kathy Sawyer, the article concentrated on the efforts that helped lead to the discovery. Our next stop was an appearance on National Public Radio's *Talk of the Nation: Science Friday*. Live from Boston, the other guest was Brian Marsden. We got into a spirited debate that nicely capsuled two opposing views of how the story should be presented to the public. This is not the kind of event that people will see from their own back yards, Brian cautioned.

I agreed that there might be little to see. Whether the public sees anything on Jupiter or not isn't the issue, I argued. The nature of the event—two solar system objects colliding—that the important thing. I suggested that people should look at Jupiter anyway, with telescopes, binoculars, even with their unaided eyes. Then they should go indoors and watch the Hubble pictures on the television. That way, people would feel a part of what was happening out there.

Talk of the Nation was a challenging program. Brian is brilliant and intense and chooses his words carefully, and I found that a successful debate with him required a higher level of alertness that I could muster that difficult day. The debate invigorated me. I returned to the Mayflower Hotel, my home for the next 10 days. Later that afternoon my mother arrived to join me for the week. It was great that she was there. I wanted so badly for Dad to be there too. As impact week neared I found that I was missing him terribly. He had died from Alzheimer's disease in 1985. He'd always encouraged and delighted in my interest in the sky, and it saddened me that I wasn't able to share this remarkable event with him. Dad loved listening to the news, every hour on the hour. I tried to imagine how much fun he would have had with the news that our comet was making.

Next on the agenda was a short drive to NASA's Washington headquarters for a series of televised interviews on a series of local stations across the United States. Although the new building was only a few blocks from our hotel, our taxi driver had no idea where we were going. There was no computer, no central place to call for directions, and other drivers had no idea what "Nasaw" meant. He finally tried to let us off at the old NASA building; then we stopped at a hotel, got directions, and finally made it a few minutes before the start of the first televised exchange. Don Savage, public affairs officer for NASA's Office of Space Science, whisked us to the studio, where I sat alone in a chair with the space agency's original symbol, a satellite rushing around the globe, behind me. Designed when NASA first began in the late 1950s, the symbol was recently resurrected by Dan Goldin, NASA's new administrator.

Then the fun began. Each 10 minutes NASA connected me to a different television studio. Some interviews were live; others were taped. Although the questions were similar, they were different enough that I found each exchange fresh and exhilarating. I was surprised at how well coordinated the three-hour program was. Don warned me that the questions might seem repetitive, but this was necessary because the viewing public changed every 10 minutes. About halfway through, I suddenly got thirsty just as someone asked why a comet impact on Jupiter was important to Earth. The series of many comet impacts had possibly brought the Earth its water supply, I explained. "Excuse me for a minute," I added as I raised my drink, "while I have a cup of comet."

These interviews were sheer joy, and they took my mind away from the distinct possibility that the impact itself could indeed fizzle. The interviews that afternoon reached a combined audience of more than seven million viewers. The impending crash already was the biggest solar system story since Comet Kohoutek more than 20 years earlier—not a great omen. I couldn't believe how long the name Kohoutek had survived in the public consciousness. Actually, that comet performed quite well, and many people viewed it through binoculars. But it wasn't the comet of the century, and because it didn't live up to expectations, to the public at large, that comet was a fizzle. Although I wouldn't have admitted this to anyone, I found myself fervently hoping that Shoemaker–Levy 9 would not be "another Kohoutek."

The broadcasting experience at NASA headquarters was instructive and pleasant, but this incredibly long day was not over yet. As George, Mother, and I climbed into yet another taxi, we set our sights toward a new studio and another kind of interview. *Petrie in Prime* was an hour-long Canadian television program anchored by Calgary-based Ann Petrie. Her own questions and insights were augmented by those from viewers calling from across Canada. Although I was enjoying this program, waves of fatigue were beginning to wash over me. But then, one caller introduced himself as a member of my high school history class. What a treat to hear, at this crucial juncture of my life, a voice from a simpler time.

On the way back to our hotel, we were stopped by a traffic jam in front of the White House. President Clinton had just announced that Israel and Jordan had reached a peace accord, but a large group was protesting the development. We finally made it back to the Mayflower Hotel, and I felt some empathy for those pilgrims of long ago, being tossed around from pillar to post. This day, which had begun years ago in a different city, was ending.

Five hundred million miles away, the grand procession of comets was closing in on its target. As the Hubble Space Telescope looked on, fragment Q was showing strong signs of elongating. Bad news, one scientist said. The piece was falling apart already! Not bad news, another countered. It was just the dust around the nucleus that was stretching out; the nucleus itself was still intact. "This change in morphology," wrote Hal Weaver for the IAU circulars, "appears to be associated with a stretching of the coma rather than a fragmentation of the nucleus, because the HST images still show sharp cores much like what we have always seen."[7] I hoped Weaver was right.

South of Beersheba in Israel's Negev desert, Jim Scotti woke from his sleep. Although his hosts were treating him well, he had an uneasy feeling. Though successfully attached to the telescope, his coronagraph was still giving some problems. "It's been hard work with lots of problems here," he e-mailed me. "The latest problem is the weather—we were all set to observe last night with the last of our major problems possibly licked when the clouds rolled in on us! Tonight is the night! Cross your fingers."[8]

New Zealand's Mount John, where Bob Marcialis was set up,

was also suffering cloudy weather. "Totally overcast—it looks grim." That wasn't Bob's only problem; a communication glitch kept him briefly out of touch with the other members of CINE. "Between the bad weather cutting out my practice on the telescope, and the isolation from e-mail, I'm feeling shaky and lacking in confidence about this whole experiment."[9] Cloudy weather was making observers nervous in England, too, where a poor weather forecast was demoralizing the observers. In Calar Alto, Spain, a team at the 3.5-meter telescope was fine-tuning its instruments. Chodas and Yeomans' latest predictions were very favorable for this site. If any effect of the first crash persisted for the 20 minutes or so that Jupiter needed to rotate the impact sites Earthward, the observers at Calar Alto would have a front row seat. Although few suspected that the fireball from tiny fragment A would be visible at all, the astronomers decided to look at Jupiter's edge at the right time—just in case.

On the international date line, the Sun was rising on Saturday, July 16. Near San José, California, Michael Liu, a University of California at Berkeley graduate student, had driven one of the dizziest roads in the country, twisting and turning its way up to the summit of Mount Hamilton and its 120-inch telescope. The University of Arizona's Don Hunten and Ann Sprague were preparing to take off aboard the Kuiper Airborne Observatory, hoping to detect, among other things, the presence of water, either in the comet or in Jupiter. At the Adler Planetarium in Chicago, astronomer Elizabeth Roettger was making a final precrash check of her e-mail. She would be a important source both for the local press and for the many visitors expected to descend on the planetarium in the coming week. In Miami, Florida, Donald Parker was taking two weeks off from his work as an anesthesiologist to image Jupiter with his telescope and CCD.

"The uncertainty only adds to the fascination," editorialized the *New York Times* on July 16. "Who can help but marvel when a comet set in motion billions of years ago reaches a long-ordained collision with our largest planet? It's enough to make mere mortals seem puny and temporary."[10]

Telescopes and observers were all in place, their tests completed and their giant optics all set to turn to Jupiter. In space, the eyes of half a dozen spacecraft were also getting ready to train their sights on the massive planet. At the Space Telescope Science Institute, the last antennas were going up as a horde of broadcasters descended on the facility. Similar crowds were gathering in places as far afield as the European Southern Observatory, whose big scopes were attracting a throng of reporters. The performers were ready, and the worldwide audience was sitting anxiously. The curtain was rising on Shoemaker–Levy 9's final act.

Armageddon at Jupiter

July 16, 1994, was the day that Comet Shoemaker–Levy 9 left the solar system and entered history. But I woke up that morning in my Washington hotel room, not to the sound of trumpets but to the ticking of my old watch. I saw the date: "16." So this is *the* day, I thought sleepily. On this day we would find out about that comet which had consumed our lives for the past 15 months. I fielded two phone calls before I was even in the shower—one from a reporter and the other from a young Montreal hairdresser who wished to discuss a series of dreams she had about crashing comets. I gave her points for persistence at least; she somehow found out where and how to reach me. She wasn't the only one who was dreaming about comets these days. I hung up the phone and thought of T.S. Eliot's *The Hollow Men*:

> *This is the way the world ends*
> *This is the way the world ends*
> *This is the way the world ends*
> *Not with a bang but a whimper.*[1]

I hoped that S–L 9's world would not end that way. As the day of first impact began, at least we knew this much: The plight of this comet had captured the public imagination. I recalled Carolyn's words during our last observing run, after I had told her of my concerns that Jupiter might just swallow the comets with no visible effect whatsoever. "Maybe that will happen," she mused. "But remember how much we've learned already. And whatever happens, Shoemaker–Levy 9 will always be remembered as the first time humanity saw a comet hit a planet."

How bright S–L 9 had been when we found it! I found myself wishing that we had discovered the comet a week after its breakup, when it would have been brighter still, perhaps bright enough to be seen through a small telescope—at least by someone familiar with comets. What a sight it would have been then! But on this day, the fragments were not really visible at all through any but

A large public gathering took place at Houston's George Observatory in Texas, on July 16, 1994, the date of the first impact. Many people, and a few alligators, attended. Photograph courtesy George Observatory.

the largest Earthbound telescopes, and only faintly through the Hubble Space Telescope out in its orbit. The comet had faded so much over the last year that ending its life with a whimper seemed a possibility. I still had faith in it, though. I still thought that the largest fragments might be 3 or more kilometers across and that they would hold together and produce a big show. That was how my gut felt, but I had no idea if that would really be what happened. From Tucson's Planetary Science Institute, my friend Clark Chapman was optimistic that, with all the firepower aimed at Jupiter, we were bound to learn a lot about the nature of comets. We would also get an insight about Jupiter's atmosphere. And we would finally learn what happens when big impacts take place.

In Baltimore, an hour's drive from the Mayflower Hotel, the Hubble science team was getting ready for their day and clustering in the data analysis area of the Space Telescope Science Institute (STScI.) Now less than four hours from impact, the doomed fragment A was on everyone's minds, especially that of team leader Heidi Hammel from the Massachusetts Institute of Technology. She was so animated on the week's television newscasts that she became a living recruitment poster for astronomy.

In England, where it was already late afternoon, clouds were gathering, a keen disappointment to observers hoping to see if any changes were visible on Jupiter after the first collision.

Never before in human history had so much attention been paid at one moment to one astronomical event. In orbit around the Earth, astronauts aboard shuttle Columbia were also looking toward Jupiter. Because they had nothing stronger than binoculars, they weren't expecting to see anything, but they watched, just in case. Amateur astronomers looked through their telescopes, and as twilight moved westward, people raised their eyes to the west, where Jupiter hung in the evening sky.

As George Cruys, Steve O'Meara, Mother, and I drove along the Baltimore–Washington parkway we picked up some news on the radio, but there was little to do now but wait. We arrived at the Space Telescope Science Institute to be greeted by a security guard who asked to see our passes. Steve O'Meara, *Sky & Telescope*, okay;

The comet's co-discoverers at their first press conference, July 16, 1994, at the Space Telescope Science Institute. Left to right: Don Savage, Gene and Carolyn Shoemaker, and David Levy. Photo by Peter Jedicke.

George Cruys, The Nature Company, okay; Edith Levy, physician, and David H. Levy—no, the guard said, there is no one by that name cleared on this list. We tried to explain that I should really be on that list, but there was no persuading the guard. Finally we persuaded the guard to radio to someone. "We have a physician named Edith P. Levy," she said, "but this David Levy, who claims to be her son, wants to come in too, and he isn't on the list."

There was a delay for a few seconds. Then a distant supervisor laughed, "He's too high on the list to be on the list. Let him through."

Back in Tucson, Clark Chapman had driven up to Kitt Peak National Observatory, where a group from the Planetary Science Institute were observing the impacts using the 84-inch telescope. Over lunch, they learned that all the open telescopes were being used for impact observations. The major aim for the night was to observe impact B in wavelengths close to what the human eye sees.

It was 4 p.m. in Baltimore. The Space Telescope Science Institute was jammed with reporters, and television cameras were following every move, not only of Gene, Carolyn, and me but of the other scientists involved with Hubble observations. It was a difficult few hours. I almost wished that I could be on the outside looking in, just to get a sense of what other people were seeing.

Now racing along at 60 kilometers per second (or 130,000 miles per hour), fragment A was encountering the outer layers of Jupiter's stratosphere. The Galileo spacecraft was not in touch at that moment with NASA's Deep Space Network. Spacecraft are not supposed to be doing anything when out of touch with the DSN; in case something untoward happened to the craft in the

The Hubble Space Telescope Science Team is a mass of excitement as the first impact pictures appear on the computer screen. NASA photo.

midst of some maneuver, engineers wanted a record of what happened. The Hubble Space Telescope had to keep its knowledge to itself until its data could be beamed back to Earth. At the Space Telescope Science Institute, Carolyn noticed herself pausing for a moment. "I shed a tear for fragment A," she said. "I have had this image in my mind of this beautiful string of pearls, and now one of them is just gone! Never again will the string be the same."

And never again would public interest be the same, I feared, if A did not put on a performance. It is the first, but fragment A is also one of the smallest pieces. Better to wait till Monday morning, Brian Marsden cautioned reporters. Although he didn't expect drama from any of the fragments, Brian felt that bright fragment G had the best chance to pack a punch. With so many warnings about a fizzle, I had even begun to trim my own expectations. If all we got was a flash captured by Galileo and a little spot by Hubble, I thought, then we'd be doing okay.

Then we heard a report about a tenfold increase in radio emissions from Jupiter, apparently the result of some event having occurred in the last day. What did that mean? As the comet's leading wing hit the planet, did a series of larger-than-expected pieces cause the radio emission to increase? We didn't know whether or not these data were significant, but at 5 p.m. CNN released this information to the world.

Meanwhile, Hal Weaver, leader of the science team observing the comet fragments as they approached Jupiter, was about to send a message to Brian Marsden about the meaning of HST's latest survey of fragment Q. He felt that a previous IAU circular contained material that needed clarifying: "I feel that it is urgent to send this info out as an IAUC," he wrote to Brian. Hal Weaver wanted to emphasize that "the decrease in the core brightness [the brightness of the inner core of the comet] could very well be due to a decrease in the dust content of the core, as opposed to a decrease in the brightness in the unresolved point source [the actual comet nucleus]."[2] Translation: Hal saw no evidence that the comet nuclei were breaking apart.

At about this same time Hal heard some astonishing news.

A few hours after Comet Shoemaker–Levy 9 swung by Jupiter in 1994, its daughter fragments might have looked like this. As the fragments moved away from Jupiter, they have already spawned their own tails. Their new lives would be short; they would return to collide with Jupiter in only two years. Painting by James V. Scotti.

This painting shows the impacts in progress as seen from an observer standing on part of the Q-complex. Fragment Q1, with its dramatic coma of dust, is to the lower left. The comet-like image at center is fragment L, and fragment K is striking Jupiter at this moment. Earlier crashes have left the line of dark spots in Jupiter's southern hemisphere. Painting by Jerry Armstrong. Courtesy Black Mountain Astronomy.

The comet impact begins in this computer-generated graphic using Hubble Space Telescope images. Courtesy D. Seale, Jet Propulsion Laboratory.

This beautiful infrared view is a false-color (or computer generated color) composite of three infrared images taken on July 18, 1994. Using the 4-meter reflector at Chile's Cerro Tololo Interamerican Observatory, astronomers imaged Jupiter three times between 2:06 and 2:18 Universal time.

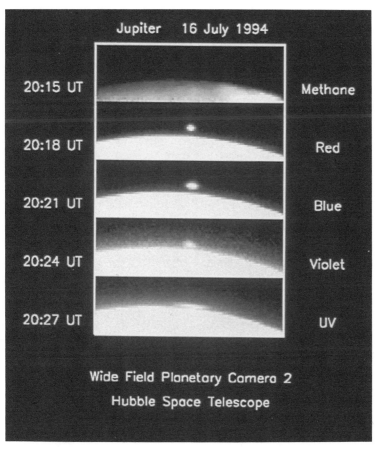

Jupiter   16 July 1994

20:15 UT	Methane
20:18 UT	Red
20:21 UT	Blue
20:24 UT	Violet
20:27 UT	UV

Wide Field Planetary Camera 2
Hubble Space Telescope

Hubble Space Telescope captured the plume from fragment A. The red image shows the plume still inside Jupiter's shadow and shining due to its own light of incandescent gas. As the plume was carried into sunlight in the lower images, condensing gases and tiny particles of dust reflect sunlight, so the plume brightened considerably. In the bottom image, the plume has cooled and collapsed to a "pancake" of material that would soon be visible as a dark spot high in the planet's atmosphere. Courtesy Hubble Space Telescope Comet Team and NASA.

"One of the STScI staff members in a nearby office," Hal recalled, "had a radio tuned to a distant station, perhaps the BBC. There was an announcement saying that something associated with the A impact had been seen at the Calar Alto Observatory in Spain. This was our first hint that this night was going to be special."[3]

But was this really the best news? "In a sense," Hal cautioned, "this news from Calar Alto only heightened my anxiety because I realized that these were infrared observations. There was a preliminary report from an optical observatory in the Canary Islands that nothing was seen. This could be our biggest nightmare: If the impacts were infrared-only phenomena, then we would be stuck trying to explain why the most expensive astronomical instrument ever built, the HST, couldn't see what even a moderate-sized ground-based observatory picked up! I couldn't let myself feel too good about these preliminary reports until I knew how HST had done, and I wouldn't know that for at least three more hours."[4]

Gene, Carolyn, and I were now up on the stage of the Institute's auditorium, preparing for the first press conference of the week.[5] Our families were already in the room with us. My mother was there; so was Gene and Carolyn's daughter Linda with her husband Fred. We were being prepped to anticipate the likely questions and work out sound bites. The idea was that Carolyn would describe the discovery, and I would follow with some thoughts about why impact week was so important. Gene would then wind up with his summary of what impact week might bring us. Each of us sat in an assigned place. The news about the radio observations put us in good spirits, but although fragment A was gone, we still did not know what, if anything, anybody saw.

Then the door suddenly opened at the back of the hall. Hal Weaver walked in quickly and whispered something to Gene. Gene stared at Hal, his eyes almost leaving their sockets. *"You mean they saw a plume!"* he exclaimed. That took care of the press conference homework. Like a comet with our own tail of reporters and cameras, we raced to a nearby office to call Brian. We wanted to ensure that he would publish a circular before 7:30 that evening, so that we could announce that a telescope near the Spanish village of Calar Alto, Spain, had spotted a plume from fragment A! We'd spent the past weeks holding down our expectations. This was better than we'd dared hope. Brian confirmed the report and added that observers at La Silla, Chile, had seen it too. In two different wavelengths, it was confirmed. "Oh, Gene!" Carolyn cried as she embraced her husband.

We walked down the hall, and I passed a smiling Miles O'Brien from CNN. "Plumes from Spain!" he said. "And Chile too," I added. "Brian is putting out a circular, and then we can announce it at the press conference."

"Hey, we've already announced it!" Miles replied. "It's on all our news shows worldwide." Surprised that the whole world knew about Calar Alto just a few minutes after I did, I hunted down a terminal and logged in to my e-mail account. With Gene and Carolyn and a crowd of people looking on, we read the historic message ourselves, the words from Spain that quelled all our doubts, and answered some of our questions:

From: mantel@USM.UNI-MUENCHEN.DE
Date: Sat, 16 Jul 1994 22:28:15 +0100
To: c1993e@astro.umd.edu
Subject: Impact A observed at Jupiter
Content-Length: 350

Information from Calar Alto

Impact A was observed with the 3.5m telescope at Calar Alto using the MAGIC camera. The plume appeared at about nominal position over the limb at around 20:18 UT. It was observed in 2.3 micron methane band filter brighter than Io.

Tom Herbst, Doug Hamilton, Jose Ortiz, Hermann Boehnhardt, Karlheinz Mantel, Alex Fiedler[6]

But there was more in my electronic mailbox that evening. We also read a confirmation message, the second sighting—in a different filter or wavelength—that proved that it was not just a dream:

The comet's discovery team prepares for a live broadcast, via Philadelphia station WHYY, over the Public Broadcasting System. Fresh from a formal dinner, we were quite uncomfortable dressed this way under the hot lights. Photo by Peter Jedicke.

> *To: sl9exploder@astro.umd.edu*
> *From: Richard Hook <rhook@eso.org>*
> *Subject: Plume observation from ESO La Silla at 10um*
> *Date: Sun, 17 Jul 1994 00:17:17 +0200*
> *Sender: rhook@eso.org*
> *Status: RO*
> *We have information from La Silla that a plume seems to have been observed at 10um with TIMMI at the 3.6m telescope. The observers are reducing the data and an image is expected to be available on the ESO WWW portal (URL: http://www.hq.eso.org/educnpubrelns/comet.html) in the near future along with further details.*
> *Richard West[7]*

What an amazing message! Not only was Richard West, one of the best known figures in cometary astronomy of the past generation, announcing the detection of a major explosive plume on Jupiter, but he also was spearheading a whole new way of doing science. By suggesting that the actual image would be released publicly over computer lines within hours, his observatory was bypassing the normal process of peer review and publication that would have delayed the publication of a picture for months.

This whole week was exceptional. "The dramatic collision of [S–L 9's] many fragments with the giant planet Jupiter during six hectic days in 1994," West later wrote, "will pass into the annals of astronomy as one of the most incredible events ever predicted and witnessed by members of this profession. And never before has a remote astronomical event been so actively covered by the media on behalf of such a large and interested public."[8] Our first Hubble photograph of S–L 9 from July of 1993 was not released until the middle of October. But this first night's images were released to the press and posted on the Internet within a few hours after they were taken. And they started an avalanche. By the end of impact week, hundreds of different images would be downloaded more than a million times from the Internet.

NIGHT OF FRAGMENT A

The crash of fragment A was an extraordinary event all by itself. That was the first time that people had witnessed such a collision. Had A been the only impact, we would have rejoiced and studied its effects for years. But in the opera of impact week, A was just the overture. One of the smallest nuclei, A was hitting the planet at a place farther away from Jupiter's daylight side than any of the others, so few scientists—Gene Shoemaker was one—expected to see the plume of material thrown into the atmosphere by its strike.

By the time our press conference began at 7:30, the reporters present were already buzzing with the news from Spain. Don Savage introduced the three of us and announced that the first

Hubble Space Telescope results would be released at a second conference to be held at 10 p.m. We are here to start the reporting on a campaign that has been 16 months in preparation," Gene noted. He cautioned the reporters that the details of the two observations from half a world away were still sketchy. But if these observations were any indication, he went on, we would be in for a great week.

As we talked and answered questions from the press, one flight downstairs the Hubble science team was anxiously awaiting the first picture from space. More than three hours had passed since the loss of A, but HST's download was delayed for an orbit because of a conflict with telemetry from the shuttle, then in orbit around the Earth. At any other time, no one would have minded a two-hour delay, but right now the wait really frayed the nerves. Finally, the space telescope was ready to transmit. A hushed crowd of scientists gaped breathlessly at the computer workstation: Hal Weaver described the scene in the observation support system room in the basement of the Space Telescope Science Institute:

"Heidi Hammel sat directly in front of the monitor where the images were to be displayed. The rest of the team formed a circle looking over her shoulder. The first image of Jupiter (taken through a methane filter centered near 889 nm) showed a hint of something near the limb, but we thought that this could be an image artifact, like a cosmic-ray event. The second image, taken three minutes after the first, didn't seem to have anything unusual. But then the third image (again taken three minutes after the previous one; due to instrument overhead, consecutive images must be taken at least three minutes apart) looked very strange; there seemed to be a little ball slightly offset from Jupiter. Could this be a Galilean satellite? I quickly grabbed a nearby copy of the *Astronomical Almanac* and, with Melissa McGrath, verified that the ball could not be a satellite.

"After the next image the answer was clear; we were watching the development of a plume off the edge of the planet!"[9]

For a moment the group of scientists just sat there, stunned.

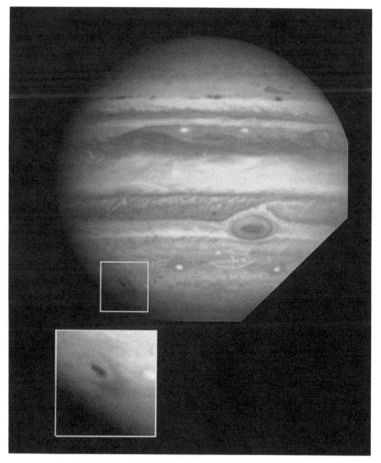

Hubble Space Telescope's first view of the spot left by fragment A after it collided into Jupiter on July 16, 1994.

There was a collective gasp. It took a few more seconds before the scientists began to realize what a treasure they had, that in this one picture, all the months of planning had paid off handsomely. "Oh," Heidi Hammel exclaimed, "My God!"

The whole picture was clear now, and the room erupted with cheering and applause. "We realized that we had something truly spectacular on our hands," Hal Weaver continued. "Melissa McGrath ran upstairs to get the champagne that she had bought for the occasion (even though she is the first to admit that she hadn't really expected to see anything like this), and Heidi Hammel popped the cork.

"The feeling of elation in the OSS [observation support system] area was indescribable, and I doubt that I will ever experience anything like this again. This was not the 'Big Fizzle' that had been predicted only one week earlier, but rather the most dazzling astronomical display of the century."[10]

"We ought to take this picture upstairs to the press conference!" said another scientist. Waiting was unthinkable. "Let's make a simple laser print of this picture, and bring it upstairs *right now*."

Back in the auditorium, Gene was describing the "latest, best attempt" of his team of David Roddy and Paul Hassig to model the impact of a hypothetical 1-kilometer-diameter comet, with the density near that of ice. The comet would break up as it encountered a certain level of resistance by Jupiter's atmosphere. Although this part of the idea is similar to other models, the rest of it is not. As reporters waited for the real thing, Gene screened three-dimensional plots from computer calculations describing a plume ballooning outward to a great distance, and then collapsing back like a giant pancake on the planet. What Gene did not know then was that the Hubble had succeeded in taking a series of pictures of Jupiter's limb using different filters. The idea—and this was a long shot—was that one of these filters might succeed in catching the plume. It turned out that the space telescope sent back a series of shots that showed the plume rising, stretching out, and collapsing, almost exactly as the model predicted, except that the actual plume was even higher. "If these reports are correct," Gene crowed, "then you can expect a good show for every telescopically observed nucleus. We're going to see things, and we're going to learn a lot. That's the good news tonight."[11]

"At that time I had no conception of all that was going to ensue," added Carolyn. "No conception of how much all of us stood to learn about comets, about a planet in our solar system, just from the discovery of this comet. "I am amazed to see people from so many disciplines work together to get the most out of this."[12]

For now, the laser print of Jupiter's new spot was ready. As Heidi left the room with it, Melissa gave her the half-empty bottle of champagne. So just as our press conference was ending, Heidi appeared at the door. Carrying a piece of paper and the champagne, she paraded before the world's press. "I think we have news from Heidi Hammel!" said Gene, and he gave Heidi his chair. Not missing a beat, Don Savage introduced Heidi Hammel.

"Gene Shoemaker always said that he'd be personally astonished if HST saw nothing," Heidi began. "Well," she gushed, "he's not going to be astonished. We actually saw some amazing things." She then held up the Hubble picture. Taken with a methane-band filter, it clearly showed, as Heidi described, "a bright streak, and around the edge of the streak there's some other stuff, a new feature where none had been there before." She answered some questions and was about to leave when she remembered the champagne. As each of us took a swig from the bottle, one of the reporters muttered "there's three more impacts!" The press conference ended in a burst of laughter and applause. "I think you can all lay your worries to rest," Gene said triumphantly. "Those reports from Chile and Spain are *right*!"

Gene, Carolyn, and I could hardly wait a look at the S–L 9 image from the HST computers ourselves, and it looked much better than the printout Heidi had shown the press conference. But fragment B was getting closer and closer to its 11 p.m. Eastern Time rendezvous with Jupiter. This was one of the only times that we would have a chance to see an impact take place. Jupiter was low in the sky but still above the horizon. The 12 and 26-inch refractors of the U.S. Naval Observatory were at least an hour's drive away, back in Washington. Out in the Space Telescope Science Institute's parking lot, we looked up at Jupiter for the first time that day. Except for the Moon, that big planet was the brightest thing in the sky. For a minute we looked at the planet in silence as a sense of history gripped us.

ON TO THE OBSERVATORY

It was time to hustle. We left a group of scientists who would be up all night poring over their wealth of images and other data. Our convoy followed Gene's race car driving, and we arrived at the Naval Observatory with time to spare for a conversation with the friendly Secret Service agents who were protecting the Vice President, whose house was on the Observatory grounds. The agents were expecting us, and they seemed as excited about fragment A as we were.

Our first look on this momentous night did not show any unusual formations, since Jupiter had already rotated the spot from fragment A across the planet so that it was once again on the side facing away from Earth. But as B got closer to Jupiter, clouds got closer to us. By B's 11 p.m. strike time, Jupiter was obscured by clouds.

Back at the Space Telescope Science Institute, the evening's second press conference was underway. "This comet was NOT a dud," exclaimed Hal Weaver, "let it ring out to the rest of the world!" Even on this night, reporters wanted to know if, as one put it, "the unwashed masses, or semiwashed masses, will be able to see the results." Although one scientist estimated that the dark material might last as long as a year, another cautioned that none of it would be visible without large telescopes and special filters. "What's in this for the guy in the street?" another reporter queried. "The guy in the street," came the answer, "can be glad he doesn't live on Jupiter!" Then Heidi Hammel expounded seriously. "This is more than science," she said. "There are things whizzing around the solar system smashing into other things with huge explosions. We live in a dynamic universe, and this is a key example of the energetics that go on in it."[13]

At Cocoa Beach, Florida, the International Planetarium Society had just completed its conference for a large group of astronomy educators. This year, to celebrate the 25th anniversary of the Apollo 11 flight, the meeting was held near Cape Canaveral. But the participants were all glued to the TV screens watching incoming pictures from Jupiter. They broke away only to call their home planetaria to handle media interviews over the phone. Whenever a major event happens, the press try to bring in a local angle. In this case, with a comet train crashing into Jupiter, the local angle from San Francisco to New York would be what the city's planetarium director thought of the show and what local planet watching parties were being planned. A little carried away by the excitement, one delegate ran wildly in circles on the beach yelling "THERE'S A HOLE IN JUPITER!" Scott Young from Winnipeg's Manitoba Planetarium saw him running, screaming, waving his arms about, and looking up at Jupiter. "I tried to bring him back to Earth," Young says; "I told him how busy we're all going to be when we got back home."[14]

At Kitt Peak, the Planetary Science Institute astronomers were anxiously watching the partly cloudy sky when their telescope operator looked up from his console. "Hey," he said, "look what I have here!" On his screen was an image from the Internet, clearly showing the plume from A as seen from La Silla, Chile. "This was really exciting," Clark Chapman recalled. "We knew now that people weren't just attaching hyped words to a minor detection. Something really dramatic was happening on Jupiter."

By 1 a.m. the day was over for us. The image of the cheering scientists was replayed over and over again in our minds and on TV screens around the world. The next day, in fact, one news broadcast showed a stadium full of people on their feet and cheering wildly. "No," the reporter said, "that was not a view of cheering astronomers. That was the crowd cheering Brazil for winning the World Cup."

At Kitt Peak, where it was only 11 p.m., Clark Chapman was about to wind down his observing. Jupiter was low in the sky, the projected impact time for fragment B was long passed. Then Clark answered the ringing phone. "Is this Kitt Peak Observatory?" a hopeful voice asked at the other end of the phone. It was a reporter from CNN. Was anybody there observing the comet crash? Great! Could you stay there and appear live, on radio, at 1 a.m.?

"No way!" Clark answered, since he was planning to return to Tucson that night. With some reluctance, the network agreed to tape him earlier. But Clark was puzzled. Fragment A's display was a brilliant one, but message after message recorded a "nondetection" of anything from B, a fragment that had always appeared as bright, or even a little brighter, than A.

Far to the west, atop Hawaii's Mauna Kea, a group of observers led by Imke de Pater did have a clear sky and the largest telescope on Earth—the 10-meter-diameter reflector that had first seen starlight just few years ago. Attached to this mighty telescope was a powerful CCD and an infrared filter. Their setup was good enough to record the weak performance of the second fragment. "We observed impact B in a narrow band L band (3.27–3.44 micron);" she wrote, "the plume was faint, but clearly detected at the expected position, starting at 02:56, fading at around 3:13."[15] B's show lasted all of 17 minutes—more than a quarter of an hour—but was far less showy than A's plume was. A quick look at the S–L 9's train gave us the possible explanation. B may have been a loosely held together group of dust and boulders. Rather than being a solid fragment, it could have been the remains of an eruption from C.[16] Even the less spectacular crashes like this one had a lot to teach us.

The B impact also revealed some strange effects through the eye of another big eye, the U.K.'s Infrared Telescope atop Mauna Kea. "Monitoring the H3+ ionospheric lines at 3.5 microns with CGS4 echelle spectrometer on UKIRT, Mauna Kea, Hawaii," observer Steve Miller wrote, "we saw a fivefold brightening of the emission around the time of impact of fragment B (around 2:50 UT) on the east limb of Jupiter. The spectrometer slit was approximately aligned on the nominal impact latitude. This faded over 90 minutes."[17] At Mount John in New Zealand, clouds were still preventing observations. "News of radio emissions and fragment A came down throughout the day," Bob Marcialis wrote. "Except for the weather, I feel encouraged. Thar's fish in the lake!"[18]

The crash of Shoemaker–Levy 9's first two fragments left the astronomical community stunned and the public interest higher than ever. In less than a day, our world had been witness to an extraordinary demonstration of power in the universe. It was as if Nature had called over the phone and said, "I'm going to drop 21 comets on Jupiter at 134,000 miles an hour.... All I want you to do is watch." As we watched with everything we had, Nature winked.

Fragments Vaulting Like Rebounding Hail

With the comet train pelting Jupiter, what kept going through my mind was the opening of Kubla Khan, the poem that had bubbled up out of Samuel Taylor Coleridge's fertile, opium-drenched mind in 1797 and which had become a favorite of my 8-year-old cousin, Emi Levy:

> *In Xanadu did Kubla Khan*
> *A stately pleasure dome decree:*
> *Where Alph, the sacred river, ran*
> *Through caverns measureless to man*
> *Down to a sunless sea....*
> *And from this chasm, with ceaseless turmoil seething,*
> *As if this earth in fast thick pants were breathing,*
> *A mighty fountain momently was forced:*
> *Amid whose swift half-intermitted burst*
> *Huge fragments vaulted like rebounding hail,*
> *Or chaffy grain beneath the thresher's flail:*
> *And 'mid these dancing rocks at once and ever*
> *It flung up momently the sacred river.*[1]

How these rocks tumbling into Jupiter were dancing! By the middle of Sunday, July 17, Jupiter was showing off scars from four collisions. When Heidi, Gene, Carolyn, and I took our places at Goddard Space Flight Center for the *Daily Comet Update*, we looked like we had stayed up at some grand and wild party. "We had an absolutely incredible night last night!" Gene began. "We all literally hit the ceiling," Carolyn added. "It was too exciting to believe, and something we will always remember."[2] We were trying to get the public, and ourselves, up to speed on some basic items, like why some images show the new Jovian spots looking dark, while others show the same spots looking bright. In the atmosphere of Jupiter, methane absorbs light efficiently, Heidi explained, so bright features, such as the impact clouds, seen at the wavelength where methane absorbs light, must be very high in Jupiter's atmosphere, above almost all its methane. But seen through filters corresponding to colors we can see with the eye, the spots appear dark because the material of the clouds is highly absorbing at visible wavelengths.

These press conferences gave us an opportunity to take a careful look at what observers were seeing all over the world. Thus I learned that at Israel's Wise Observatory, Jim Scotti's big telescope was surrounded by several amateur astronomers looking through small telescopes. Two of these observers reported seeing a flash just as A came down. Whether these observations could be confirmed was an unanswered question, but it was definitely time to issue a red alert for amateurs to watch for possible flashes from future impacts. Apparently no one captured these flashes on film.

The origin of the large semicircular area circling the center of the A impact site was a mystery that Sunday. Gene was sure that it was the atmospheric wave, a visible manifestation of a wave heading through Jupiter. "NOT!" Heidi insisted. "It is debris falling onto the ammonia clouds." After some arm waving, Gene and Heidi agreed that time would tell who was right. Time did; Heidi was.

Sunday's press conference was useful because it helped de-

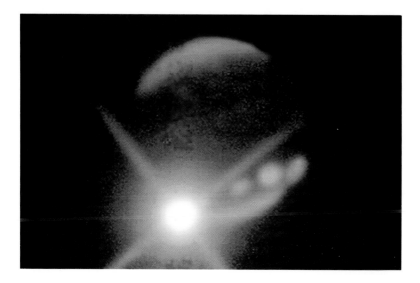

Jupiter as diamond ring: On July 19, 1994, Peter McGregor and Mark Allen used a 2.3-meter reflector to record the bright plume left by the explosion of fragment K. Although the spikes are an artifact caused by the telescope's secondary mirror, they add to the drama of the picture. Courtesy Peter McGregor and Mark Allen, Australian National University.

Appearing on the planet's limb is the first of several hits from what once was fragment Q. This image was taken through the 3.5-meter reflector at Calar Alto, Spain. Courtesy Tom Herbst, Max Planck Institute for Astronomy.

Shoemaker–Levy 9 meets Jupiter. Painting by William K. Hartmann.

A fragment of Shoemaker–Levy 9 impacts Jupiter in this dramatic painting by artist Don Davis.

fine the event in terms that the public could understand. For example, it marked the death knell of the confusing use of both letters and numbers for identifying fragments. The IAU circulars had identified the fragments both by letters, with A being the first fragment to come in, and by numbers. At Sunday's press conference, a graphic appeared with both letters and numbers shown. "Get rid of the numbers on the graphic!" Gene asked, and a minute later a new graphic appeared with only letters. Although the numbers were still used along with letters on the IAU circulars until the end of the impacts, I did not see them used after that. The press conference also helped to clarify why Jupiter was being bombarded so badly. More than once Gene explained that the energy released in a collision depended on how fast the comet is traveling. There is a big difference between a rock dropped into a pond and the same rock shot from a cannon into the pond. The energy increases as the square of the velocity. At 60 kilometers per second a lot of energy was being released, even from fragments of modest size.

Then came a question about the visual observations of the flashes from the first impacts. Even though these observations were reported from experienced visual observers, all four of us on the panel were skeptical. "When you really want to see a flash, you can almost imagine that you do," Carolyn correctly explained. Heidi added that the observations had to be doubted because they were not confirmed by any of the many professional telescopes, "and they were watching for this." Because I was the amateur in the group, a few other amateurs came down hard on me for not rushing to their defense. "He called all amateur astronomers liars!" exaggerated one. What I did say was that this is a red alert for all visual observers who can observe Jupiter during future impacts. If the flashes are that easy to observe, other visual observers will see later ones. My new public status had its drawbacks, but I was surprised at the complaints from amateur astronomers, people I had thought were kindred spirits.

I was egregiously but happily wrong about another prediction I made that day. I doubted that small telescopes—2- to 4-inch-diameter telescopes that inhabit many homes—would be strong enough to see Jupiter's new dark markings. "I will be thrilled and delighted," I added, "if I am wrong on that."

Near the end of the conference, a reporter from *Earth News* asked Gene how last night's extravaganza compared with another event with which he had been deeply involved—the landing on the Moon 25 years earlier. "That's hard to compare," laughed Gene, "for they were different events. I think we all had a lot of confidence in Apollo 11—that it was going to be a successful landing. We were delighted and grateful for that achievement. What's different with this event is that we didn't know how Nature was going to perform. We're just elated that Nature has outdone herself to give us this spectacular experiment."[3]

When C released as bright a fireball, and a spot, as A did, it was easy to figure out what had happened to B. Hal Weaver expressed the idea that a number of us had: "In pondering the question of why nucleus B didn't produce as big an effect as nucleus A (can we agree to call them "nuclei" now!), the one thing that jumps out at me is that A is on the train [the straight line that most of the fragments followed] while B is significantly displaced from the train."[4] That confirmed the suspicion we'd had right after B hit—that the second fragment had been a loose bundle of boulders, possibly spawned by an eruption from C.

"The first night proved the whole thing wasn't a big dud," wrote Michael Liu, the Berkeley graduate student who was observing the impacts from northern California's Lick Observatory. "Tonight proved it wasn't a one-hit wonder."[5]

Startling as the first impact was, what seemed just as exciting was that each of the hits so far was different; each held its own surprise. From the Australian National University's 2.3-meter telescope, near Coonabarabran, New South Wales, Peter McGregor saw two impacts associated with fragment C. Separated by about an hour, the second was far brighter than the first. "This event brightened appreciably during the first five minutes," he wrote, "and then faded to the brightness of the first event after 10 minutes. Both features are still visible one hour after the initial impact."[6]

The Hubble Space Telescope records fragments C, A, and E in violet light on July 17, 1994. Courtesy Hubble Space Telescope Comet Team and NASA.

This was the first indication of impact clutter, a name I give to a condition that would quickly become more serious as more comets bombarded Jupiter. "It seems likely," wrote observer Mike Brown, "that the 'early' fragment C impact observation at 06:24 UT is actually the first reported observation of the reappearance of the fragment A impact site after one rotation."[7] Translation: Fragment C did not hit early. What was thought to be C was really A returning as the planet went through a full 10-hour rotation. This was special news in itself. Few people suspected that any of Jupiter's wounds would remain so visible for a whole 10-hour rotation of Jupiter.

As C's large spot—bright when observed through a methane

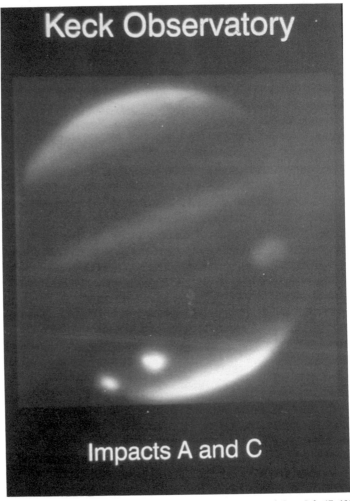

*This infrared view from the Keck Telescope shows fragments A and C on July 17, 1994.
Courtesy I de Pater.*

Donald Parker, an anesthesiologist living in Coral Gables, near Miami, Florida, recorded these stunning views of Jupiter using a 16-inch telescope and CCD from his home in Coral Gables, Florida. The series of Parker images in this book show incredible detail and are among the finest views of Jupiter ever taken by anyone, amateur or professional.

filter, but dark at visible wavelengths—approached Jupiter's western limb, a new spot appeared as D tumbled into the planet. From Australia, Peter McGregor observed a short-lived fireball, in infrared light, half the brightness of Jupiter's moon Europa.[8]

A WHOLE NEW WAY OF DOING SCIENCE

The excitement of the impacts powered some of this astonishment, but it didn't provide all the fuel. Emerging in the surge of events was a revolutionary way of sharing scientific results within minutes, not months. The ability to share words and images instantly, using what computer types call an e-mail exploder, had

This composite of Jupiter was prepared from three different infrared images taken on July 18, 1994 using one of the southern hemisphere's largest telescopes, the 4-meter diameter reflector at Chile's Cerro Tololo Inter-American Observatory. The bright spot left from fragment A is seen at lower left, and "tiny" D shows up well at right. Courtesy John R. Spencer, Darren DePoy, Jay Frogel, and Nicholas M. Schneider.

welded the world of planetary astronomers merged into one big observing team, even though members were using telescopes scattered all over the Earth. Information sent to the "SL9 Message Center" at the University of Maryland was immediately relayed, using the Internet, to observers and theoreticians in countries around the globe. This way, observations were reported and initial interpretations made and made public on the fly—as opposed to days later in memos and months later in journals. Afterward, some of the observations and even more of the interpretations were recanted or altered, but the instant exchange was widely applauded as a healthy development in a scientific discipline where observations are not always released quickly. Meantime the large community of amateur astronomers was chiming in with their reports of sightings. Many amateurs tuned in via computer to one of the big electronic bulletin boards, like CompuServe, Genie, or America On-Line. *Sky and Telescope*, The Nature Company, and the Planetary Society were busy updating their dial-in recorded messages for thousands of people eager to receive more current and detailed information than they could get through the news media.

By the middle of Sunday afternoon, even more computer users were logging into the Internet sites that offered pictures of fragment A colliding with Jupiter. Few as they were this early in the week, the pictures were so striking that computer aficionados around the world, whether or not they had any previous interest in astronomy, were eager to see first hand what was happening out at Jupiter. By the end of crash week, more than 1.6 million people had logged on to the Jet Propulsion Laboratory's New-products computer, the largest invasion ever launched on the Internet.[9] "During the impact week," wrote Ron Baalke of Jet Propulsion Laboratory, "I was working around the clock to keep the home page current, even though I was driven to total exhaustion."[10] His heroic effort paid off. By early November his "home page," with its 800 available images, was accessed more than 2.5 million times.

My own electronic mailbox was rapidly beginning to fill up

with messages from around the world. Most of this e-mail was about the A, C, and E impacts. D was a granddaughter fragment. Just as B had split off from C, fragment D had fallen away from E, and it dropped into Jupiter with a relatively mild splat.

The splat was not so minor at the University of Chicago's SPIREX (for South Pole Infrared Explorer) site at the South Pole. Atop a 10,000-foot-high plateau, a small group of observers faced extraordinarily tough conditions. In return, Nature provided them with crystal-clear conditions, virtually no humidity, and nonstop darkness with Jupiter in the sky for every impact. But they had to fight for some of their data. "The SPIREX fragment D data was significantly compromised," the hardy team reported, "due to the sudden onset of low blowing snow. The telescope was heroically cleared of accumulated snow by Joe Spang and John Briggs … in strong winds at temperatures of −60 degrees Celsius."[11] They completed their frigid work in time for a spectacular observation of the next car in the train.

At 11:17 Eastern Daylight Time, July 17 (15:17 UT), the first large fragment cascaded in. "E is a dilly!" Gene said excitedly. In its wake a plume of material soared upward almost 2000 miles above Jupiter's cloud tops. "Soon after the predicted time of impact of fragment E," wrote astronomer John Menzies, "a bright plume at the limb of Jupiter was seen … in the K band images. The development was similar to that of fragment A. As the impact site rotated into view, a large bright spot, about half the diameter of the Great Red Spot, gradually appeared. Soon thereafter a second spot appeared at the limb, and when it rotated farther it was seen to be centered in a dark circular patch. By about 2 p.m. (18 hours universal time) a third spot appeared at the limb, again apparently surrounded by a dark patch. Meanwhile, a dark circular feature centered near spots 1 and 2 could be seen easily in the eyepiece of the 0.75m [¾-meter] telescope at SAAO, Sutherland. Stay tuned."[12]

At Calar Alto, the observatory made famous by their premier detection of the first impact, fragment E came in, producing a bright and obvious plume, rapidly increasing "to more than 30 times the brightness of Europa." It stayed brighter than Jupiter's moon for more than six minutes.[13]

Now it was F's turn. Early reports indicated that the fragment was a disappointment, with no visible flash. However, both the big 3.6- and 2.2-meter telescopes at the European Southern Observatory did record a small plume, and some observers did report a dark spot of some material appearing near the eastern limb of Jupiter when the planet rotated the impact site into view.[14] But since F had come down almost on top of the 10-hour-old E site, "impact clutter" started again. Three messages, all within about half an hour, illustrate how severe the problem was becoming, and how useful the exploder was in helping unravel the confusion.

"The impact F site is now as big as the Great Red Spot," wrote Lowell Observatory's John Spencer, "and shows two components, one many degrees of latitude south of the other, at 2.3 microns [the methane band]. Considerably brighter than the south polar cap now (at 02:10 UT). South component is fainter.

How do the impact sites get so big so fast?"[15]

Then followed a message from McDonald Observatory: "Although it is possible that this bright spot is the impact of F with a substantial time delay past the predictions, it is somewhat more likely that this is the impact spot of piece E. The surface brightness of this spot is greater than the south polar hood."[16]

Spencer quickly agreed: "The very bright and complex spot we are now observing is the old E site, not the new F site. Perhaps the complex morphology is due to the overlap of the two impact sites, though? We now are observing the A impact site coming into view at the start of its fourth rotation."[17]

By the late afternoon of July 17, the sky finally cleared over Cambridge, England, where the British Astronomical Association's John Rogers was observing with a 30-centimeter (12-inch) reflector. He noticed dark splotches as large as Jupiter's oval spot known as "BC," a whitish Jovian storm that has been observed for several decades. However, the new spots were dark and far more obvious in appearance than this feature. The spots, Rogers noted, had grown in size since they were first recorded by the Hubble Space Telescope.[18]

Meanwhile, observers in Israel reported something most puzzling: "Two distinct stretched spots were observed in the Jupiter

NORTHERN hemisphere, North North Temperate Belt, on both sides of the Central meridian, at July 17 19:05 UT." The spots were two times bigger than the dark impact site seen on the southern hemisphere, presumably from the E nucleus.[19] The K impact also produced features on the Northern, opposite, polar region. They were interpreted as unusually intense auroral activity—actually a northern light display caused by charged particles following what we could call a magnetic highway stretching from Jupiter's southern to northern regions.

Now the world waited for G. Although it was supposed to be big, a small piece had been observed to break off just a few weeks

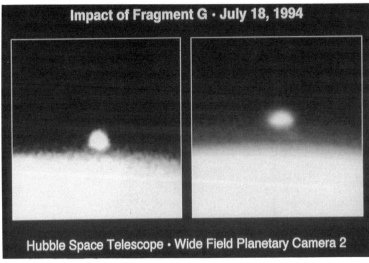

Impact of Fragment G · July 18, 1994

Hubble Space Telescope · Wide Field Planetary Camera 2

These two spectacular images of the G impact plume were taken with the Hubble Space Telescope. The plume extended to about 3300 kilometers above Jupiter's cloud tops. In the image at left, the plume is still within Jupiter's shadow, and so it is lighting itself. At right, the plume has climbed into sunlight. However, we can still see the part in shadow, fainter and toward the right; the plume is climbing at about a 45-degree angle. Courtesy Hubble Space Telescope Science Team and NASA.

earlier. Did this mean that the entire fragment would fall apart before impact? Another possibility: G was so large that it might penetrate more deeply into the planet's gases before being stopped, and not all the material would erupt into the fireball. If this occurred, G could leave a smaller plume than A had.

"MY GOD, IT WAS EXTREMELY BRIGHT!"

In the predawn hours of Monday, July 18, a tiny fragment, unseen by anyone before now, approached Jupiter and collided with it. As it went down, its small flash was seen by at least one telescope. But this was just a precursor. Thirty seconds later, the gates of hell opened as the central mass of fragment G blew up, leaving a mighty fireball soaring some 3,000 kilometers (more than 2,000 miles) above the clouds.

Back on Earth, astronomers were raving. "The plume from G was as bright as Jupiter itself!" one person wrote after seeing the image in the light of methane and 2.3 microns. At the Australian National University's telescope, Peter McGregor's CCD detector was so filled with light that the image formed a bright cross of diffraction spikes, causing an otherwise darkened Jupiter to shine like a diamond ring. It was first detected at 7:33 Universal time (UT). "The brightness of this feature increased by about a factor of 10 by 7:35 UT and remained stable until about 7:40 UT," he wrote. "At that time, the G impact site brightened enough to saturate our detectors and produce brilliant diffraction spikes. The mirror was then stopped down to 1.9 m [that means the telescope was effectively reduced in aperture] to prevent saturation." Minutes later, as Jupiter rotated the impact site to a more favorable viewing position, the Australian observers saw it clearly on the TV image of their guiding camera.[20]

From all over the world, astronomers reported incredible observations. From the South Pole: "Impact G does appear to be big, and extremely long lived. It has been bright for over 30 minutes. The words of Hien Nguyen, at the South Pole, were: 'My

The plume from fragment G as seen through the Keck 10-meter telescope on Mauna Kea, Hawaii. Courtesy I. de Pater.

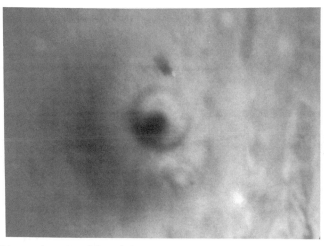

The G impact site as viewed through the Hubble Space Telescope, but computer processed so that it appears as though the telescope were directly above it.

God, it was extremely bright!' This could be an understatement."[21] From Japan: "Huge dark spot larger than red spot was continuously observed until the 'spot set' [the setting of the spot over Jupiter's western limb as seen from Earth] with optical CCD camera attached to 91 cm [about 36-inch aperture] telescope at the Okayama Astrophysical Observatory."[22] From China: "A strong radio burst from Jupiter was detected.... The signal arrived 31 minutes earlier than the predicted impact time of fragment G of SL9. We suppose that such a strong burst is due to the cyclotron emission when fragment G transpassed [entered] the magnetosphere of Jupiter which has a larger extension [extends further out] than the Jupiter atmosphere."[23] And from South Korea: "We could observe the site of the G impact at 11:28(UT) with the 1.8 meter Bohyunsan telescope. CH_4 [methane] filters in the visible range were used. The size of the G area was comparable to that of the Great Red Spot." They also reported observations of the G site

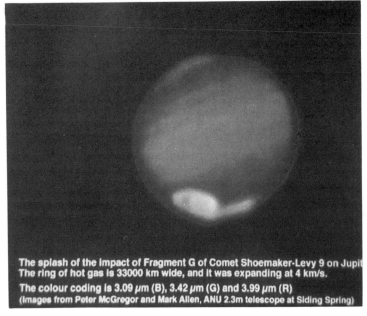

The splash of the Impact of Fragment G of Comet Shoemaker-Levy 9 on Jupit
The ring of hot gas is 33000 km wide, and it was expanding at 4 km/s.
The colour coding is 3.09 μm (B), 3.42 μm (G) and 3.99 μm (R)
(Images from Peter McGregor and Mark Allen, ANU 2.3m telescope at Siding Spring)

Impact of fragment G. The ring of hot gas, shown greenish here, is some 33,000 kilometers wide, and is expanding at 4 kilometers per second. Courtesy Peter McGregor and Mark Allen, Australian National University 2.3-meter telescope at Siding Spring.

with several other telescopes, including a 20-centimeter [8-inch], even though July is monsoon time in Korea and good observing conditions are almost nonexistent.[24]

Observers atop Hawaii's Mauna Kea had a touch-and-go bout with clouds that ended miraculously. "The summit here in Hawaii," wrote Imke de Pater, "is plagued by heavy fog. All telescopes are closed. At 21:27 HST [07:27 UT], a minute or so before the expected impact, the IRTF [the Infrared Telescope Facility, another telescope on the mountaintop] noticed a clearing, and we opened up. At 21:39 [07:39 UT] we obtained our first frame of Jupiter, in regular K-band: a truly remarkable (saturated) plume was visible well above the limb. We started a sequence of observations at 3.4 microns: a spot was visible in our first frame, 21:40 HST [07:40 UT] which brightened to truly remarkable levels by 21:50 [07:50 UT], after which it decreased in intensity. At the same time the fog was coming in, and we were closing up again; whether we were seeing a true decrease in the spot's intensity, or whether the decrease is due to increasing cloud coverage above the telescope is not yet known.

"It is raining right now; we don't expect to get any more frames tonight."[25]

The Hawaiian mountaintop must been quite a place one minute before the expected impact time, with domes suddenly opening all at once to respond to the sudden break in the clouds. Next door, at NASA's IRTF, observers were also blessed with the hole in the clouds and rain. "Looking through fog and 98% humidity, we imaged impact G at 2.29 and 4.78 $+/-$ 0.11 (M) microns [a filter emphasizing light in the infrared] with a time series of 0.9 sec integrations taken every 7 sec in alternating filters. The series began at 7:35 UT, when precipitation stopped briefly, with the impact site already bright. The impact region increased sharply in brightness … eventually saturating the detector during its brightest phase. We estimate that the impact site outshone the planet. Rapidly varying fog conditions prevented a real-time analysis of the 2.29-micron images."[26]

From Mexico's Baja Peninsula, James Jay Klavetter added a personal anecdote to his team's scientific work: "I observed the S–L 9 impacts at San Pedro Martir Observatory (SPMO) with an old MIT friend, Steve Levine, who was a postdoc at Observatorio Astronomica Nacional (OAN). We were not only friends, but mountaineering partners, and Steve said I should bring some of my climbing gear since there were some nice rocks and cliffs around.

"After a disappointing first night, but seeing and reading the reports of the A and C impacts, we took a little time off the next day to explore some of the outcroppings near the 2.1 m dome. We found a beautiful corner: steep and clean with enough challenge

to make it interesting. Steve and I both climbed it and, just as with asteroid discovery, we were entitled to naming the climb (as far as we knew, we were the first to ascend it). We decided to name it 'Fragment B' based on the exploder reports, since it was a hidden corner and no one saw our ascent.

"A couple of days later, we found time to get out again. We found a beautiful rock cliff with two obvious lines. We called the rock 'G' since it was the day after the G site was seen on Jupiter and because it was just a perfect cliff face. We climbed both lines and called the easier one 'G1' and the harder 'G2' after the exploder messages telling us there were two distinct G impacts.

"Although it will probably never get widely known, some of the fine climbs around the 2.1 m telescope at SPMO are named after S–L 9 impacts!"[27]

Out near San José, California, at Lick Observatory, graduate student Michael Liu noted that the G impact site was extremely dark—"much larger than the spots from the other impacts (A, C, E/F). The site was accompanied by a dark crescent or 'eyebrow' to the south." Later on, most astronomers would think that the eyebrow was the result of material being thrown into the Jovian atmosphere by the explosion and then falling back again. "Astronomers looking through telescope eyepieces were on every news report on every TV station. I heard an interview on NBC with an anchorwoman and an astronomer in Australia," Liu went on. "She couldn't resist asking, 'So what does this comet data mean to you personally?' The astronomer answered by saying it was the best thing that had ever happened in his life, like 'a whole load of Christmases rolled up in one.' Can't say I feel the same, but this thing has been pretty cool."[28]

Almost all astronomers reported that their observatory time was being shared with the press and the public. At Lick, the old 36-inch refractor, under whose pier the wealthy and philanthropic James Lick lies buried, hosted a class from a local community college. "Climbing the ladder," Liu writes, "I got a chance to see

The plume from the crash of fragment G towers more than 3000 kilometers (2000 miles) above the tops of Jupiter's clouds in this series of images from the Hubble Space Telescope.

This picture was taken 12 minutes after fragment G collided with Jupiter. It was taken using a 2.34 micron filter on the CASPIR instrument attached to the 2.3 meter Australian National University telescope at Siding Spring, Australia. Photo courtesy Peter McGregor.

impact site G through the eyepiece of the refractor—both the spot and crescent were very obvious. A huge thrill to see it by eye, as opposed to sitting in front of a workstation—surely a once in a lifetime opportunity. (Have I said this before?) Everyone was excited, especially the teacher who brought the class. We had a reporter from the *San Francisco Chronicle* who observed our observing tonight. We saw plenty of stories on TV however. The big

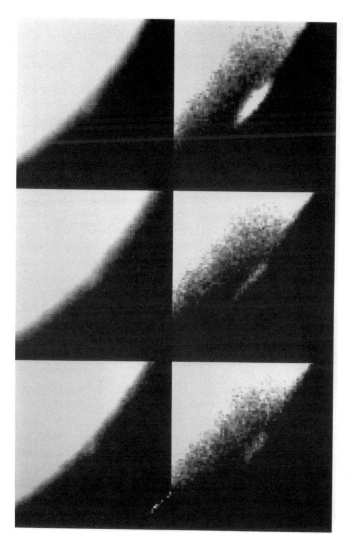

Steve Larson and colleagues used the 4.2 meter diameter William Herschel telescope to take these remarkable images of the plume from fragment H, on July 18, 1994. Images courtesy S. Larson.

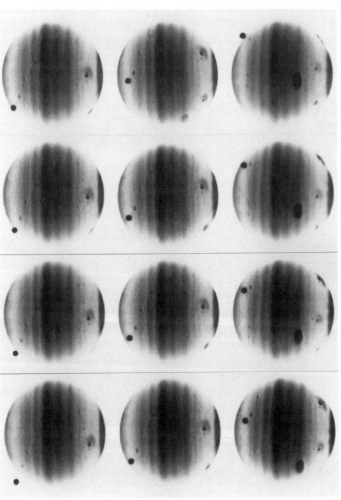

Using the 4.2-meter William Herschel Telescope, Steve Larson and colleagues captured the D–G-complex rotating across the face of Jupiter (left to right, top to bottom) on July 18, 1994. The newly formed H spot is just appearing in the second image. The moon is Ganymede. Courtesy S. Larson.

Using the 4.2-meter William Herschel Telescope, Steve Larson and colleagues captured these images in three filters. The time is the same as the image in the preceding figure, middle row, far right.

question is whether the comet was big enough news to make it into the opening monologue of Letterman."[29] (It did.)

At 41,000 feet above the Earth, an aircraft equipped with telescopes to take advantage of air thinner and clearer than on any mountaintop had its telescope trained on Jupiter. Although the Kuiper Airborne Observatory (KAO) is based at NASA's Ames Research Center near San Francisco, the big C-141 can fly anywhere in the world and during impact week was taking off and landing in Melbourne, Australia.

Scientists flying on the KAO successfully detected methane emission at 7.7 microns associated with the G fragment, but at other wavelengths failed to see any Jovian water vapor. They and other groups were starting to think that the lack of water was evidence that the comets exploded above where Jupiter's water layer was believed to be.[30]

Using telescopes as small as 30 centimeters (12 inches), amateur astronomers were flabbergasted at the ease with which the dark patch from fragment G could be seen. In Miami, Florida, anesthesiologist Donald Parker had spent weeks carefully choosing the best filters to view and image the planet, in the hope that he might see some evidence of the impacts. On the night of July 18, he looked through the telescope. Using no filter at all he could see a big black spot that was more obvious than anything he had ever noticed before on Jupiter.

Parker continued his observing each evening until, as he was climbing the ladder to his telescope to image Jupiter just after impact week, he felt a pop in his knee. He had torn some ligaments, an injury which sadly resulted in surgery, months of pain, and medical leave from his profession. Not until December was he well enough to travel to the American Geophysical Union meeting in San Francisco to display his incredible Jupiter images.

After his evening at Kitt Peak, Clark Chapman was now observing from his Tucson home. A few months earlier he had blown the dust off his childhood 10-inch-diameter telescope. The instrument hadn't been used much since 1962, and had not been used at all for almost 15 years. Its mirror was badly in need of a new coat of aluminum, so the telescope no longer had the light grasp of an new and well-maintained 10-inch.

Clark's impact plan was to use three telescopes at three locations: an 84-inch on top of Kitt Peak, a 4-inch far away in space, mounted on a spacecraft on its way to Jupiter, and the 10-inch in his backyard. He knew that the Kitt Peak program was proceeding well. Its strategy was to observe Jupiter in a series of scans, so that the planet would appear as a bar of light and impact flashes would grow and then vanish on the limb. How Galileo was doing was not yet known. The spacecraft had its instructions and seemed to be executing them on schedule, but the data playback for the images would take weeks. Clark had spent many hours devising two different imaging modes that the spacecraft would work on during different impacts. Some would be straight snapshots, but others would take the form of still movies, like at Kitt Peak; this time Jupiter would appear as a bar of light as the spacecraft's scan platform slowly moved. This way, a single exposure could tell the full story of a fragment's fall into Jupiter.

But on Sunday, July 17, Clark was in his front yard with his 10-inch telescope. Observing with his daughter Jeanette and his new wife Lynda, he peered through his telescope and was stunned to see the dark spots from fragments A and E without any trouble at all. "We couldn't believe how obvious they were," he said later. "E was as dark as the shadow of Ganymede, Jupiter's largest moon." That evening he watched a replay of the day's press conference from Maryland, in which Gene Shoemaker remarked that these spots might be viewed by amateur astronomers with large telescopes. Clark needed to set that right. He quickly wrote Gene an e-mail message that the A and E spots were clearly visible to anyone with a "modest glass" or small telescope. "Don't be conservative now, Gene!" Clark thought. Jupiter was changing fast, and almost everybody could see it.

On Monday night Clark looked again. This time he was greeted by a spot so big and black that he could not believe what he was seeing. "Waves of awe swept over me," he later recalled. Jupiter was reacting to fragment G in a very big way. Meanwhile a crowd was gathering at the Flandrau Planetarium in Tucson, where telescopes were set up to look at Jupiter. At first there was confusion, for one of Jupiter's big moons was leaving its own big shadowy mark near Jupiter's equator. Forgetting that the comet train was hitting over Jupiter's south polar region, some people thought that the satellite shadow was an impact spot.

Five hundred miles away, my close friend Steve Edberg had organized an observing session for his colleagues Paul Chodas and Don Yeomans—the people who had calculated Shoemaker–Levy 9's orbit with such precision—and members of their families. Despite the fact that this was a warm night with telescopes set up on a street, Jupiter appeared sharp and steady. "My most memorable view of Impact Week came this night," Steve writes. "G had just rotated into view and we saw it as an elongated ellipse with a bright center and a dark spot at one end!"[31]

At Mount John in New Zealand, Bob Marcialis was finally able to get some images of the G site through clouds. It was difficult to get the right exposure with clouds changing Jupiter's brightness all the time, but he did manage to get some successful images.[32] At least he was observing.

What were the Shoemakers and I doing that Monday evening? We were not observing. In cloudy Washington, we were at a dinner arranged by the Planetary Society. The evening was important in that it provided a chance for us to discuss the impacts with members of Congress who were just becoming alerted to how exciting the sky can be. But I found myself wondering why I hadn't brought along Minerva, the small 15-centimeter (6-inch) telescope I often take on the road. I had never suspected that the spots would be visible through her. But I was mistaken. Oh well, I reasoned, at least it was cloudy here in Washington.

By Tuesday morning, July 19, the surge of reports of Jupiter's dramatic appearance had reached fever pitch. The Internet was clogged with born-again skywatchers trying to see the new Jupiter. On their computers, the colors and richness of the images supposedly rivaled the original. Thousands of people now had access to images technically as good as what the Hubble Space

Telescope scientists were analyzing at that moment. So heavy was their use of Internet sites, especially the one at Caltech's Jet Propulsion Laboratory, that at peak times the information superhighway threatened to shut down from all the traffic.

Also by Tuesday morning, it had become clear that the broadcast networks and print media were devoting more and more attention to the story as fabulous pictures and reports continued to come in. The press conferences were now being carried live on some stations, so on Monday millions of people were tuning in to watch the scientists debate what was causing the tumultuous effects on Jupiter. Never before had a major astronomy story been given this much attention.

"I want to put this into the historical context of Jupiter observations," Clark Chapman wrote late Monday evening. "I have just come in from looking at Jupiter with my backyard telescope. The preceding end of impact site G is approximately on the central meridian. Based on my own extensive experience of observing Jupiter when I was younger, and studying historical records of Jupiter observations from the early drawings of Hooke and Cassini through the extensive nineteenth and twentieth century reports of the British Astronomical Association, I would assert: THIS IS THE MOST VISUALLY PROMINENT DISCRETE SPOT EVER OBSERVED ON JUPITER. (By 'prominence' I mean the combination of both size and contrasting albedo.) Does anyone disagree?"

According to Thomas Hockey, whose thesis concerned spots on Jupiter, there have been other instances of dark spots, but none as prominent as these: "There is nothing in the photographic record of Jupiter to dispute Chapman's assertion," Hockey wrote. Hockey did uncover a few examples of spots that date back to 1690, when Cassini observed a large spot in Jupiter's south equatorial belt. In 1778, William Herschel drew three large equatorial spots on Jupiter, each of which occupied less than a tenth of the planet's diameter. In 1834, George Airy described "a remarkable spot seen on the apparent southern belt, nearly four times as large as the shadow of the first satellite." In 1850, Dawes and Lassell recorded a series of spots in the South Temperate Zone, "two being nearly equal in size, and almost as large as the third satellite appears...."[33]

According to Steve O'Meara, who has studied old visual observations of the planets for many years, it is difficult to compare drawings of dark spots with modern photographs, since drawing techniques often resulted in any dark spots appearing darker than they really were.[34]

"Having last night (03:00 UT) visually observed Jupiter through the 0.2-m telescope at the Hillside Observatory," Hockey concludes, "I agree with Chapman that we are witnessing spots that would have amazed the great planetary observers of the 17th, 18th, and 19th centuries."[35]

12

*Crashes,
Congress, Clouds,
and Clothes*

"I feel sorry for Jupiter," Heidi Hammel said on the afternoon of July 18, 1994. "It's really being pummeled!"[1] The next day, the U.S. Congress responded to the battering of the punch-drunk planet. Chaired by George Brown, the House Committee on Science and Technology voted to direct NASA to mount a program to find all the objects that could present a threat to the Earth—all near-Earth asteroids a kilometer or greater in diameter, and as many comets as possible. (Since comets take much longer to orbit the Sun than asteroids, and are more rarely seen, finding all of them will be a much more daunting task.) This vote reflected the nation's fascination, and its growing awareness that such a trail of destruction could have been headed at Earth. The initiative was designed to protect future generations of people. Today's youngsters who were waking up to the fact that the Earth's neighborhood was a crowded place and that there was a chance in a thousand that something big could hit us in the next century. NASA was already far along on a proposal, called Spaceguard, that called for a series of wide-field telescopes at locations around the world. Using electronic CCDs, these telescopes would search for intruders. A commission chaired a few years ago by David Morrison, a well-known planetary scientist at NASA's Ames Research Center in California, had proposed Spaceguard. After the spectacular collisions on Jupiter, it was time for action on Earth. Gene Shoemaker was tapped to chair a new committee and report back to Congress within six months, by February 1995, on his findings.

This outgrowth of the collisions was an exciting development, and astrocelebrity Carl Sagan later said that it could have significant ramifications. If the new search program revealed that an asteroid or comet was indeed about to hit us, and if we changed its orbit in time, it could then be suggested that Shoemaker–Levy 9 rescued our planet.[2]

Congress was apparently as amazed as the rest of us about what was happening some 500 million miles out in space. And much more was happening on this day, Tuesday, July 19. From the University of Maryland, Ann Raugh was doing a superb, virtually round-the-clock job monitoring the many messages coming through the University of Maryland's e-mail exploder.

As the long-awaited K impact took place on Jupiter, observers eagerly turned their telescopes to Jupiter's ice-bound moon Europa. It was being eclipsed by Jupiter at that moment and was thus capable of reflecting the light of the impact from its darkened surface to Earth. With the eruptive plumes looming large and bright in infrared light, some observers had fully expected to see the moon brighten up significantly. Others, including Brian Marsden, had suggested that the brightenings would not been seen. S–L 9 proved Brian right: Europa did not brighten as had been expected. In fact, of all the telescopes specifically observing Europa, only one saw any magnitude increase at all! Had K simply forgotten to fall? Not quite. Minutes later, Peter McGregor in Australia detected another enormous plume as Jupiter once more looked like a diamond ring. And this time the ring encircled a darkened planet decorated by the cooler clouds from three of the earlier impacts. "The impact of fragment K was recorded in clear skies … on the ANU [Australian National University] 2.3 m telescope at Siding Spring Observatory," McGregor wrote. Earlier, and his colleagues had seen no reflection off Jupiter's satellite Europa. "The first indication of a plume was detected at UT 10:21:22 at 2.34 microns. [Three minutes later] at UT 10:24:33, a bright fireball was detected which grew to twice the size of the nearest remnant impact site by UT 10:25:54." The fireball took only 90 seconds to do this, but more was to come. The site blazed for six minutes (until UT 10:31:40), when suddenly, "an intense central core brightened sufficiently to cast diffraction spikes across the CCD array. As the impact site rotated into view, the fireball increased in brightness rapidly. By UT 10:33:02 we estimate that it was slightly brighter than the peak brightness of impact G observed yesterday." To top off this remarkable observation, the Australian team detected a wave moving out from the impact site.[3]

At long last, the New Zealand sky cleared for Bob Marcialis, and he spent his evening getting superb images of the remaining fragments, as well as spectra of the big K site as it began its march across Jupiter's face.[4]

At the *Daily Comet Update* on July 20, the University of Arizona's Roger Yelle announced a major discovery: Using Hubble's faint object spectrograph, he and his colleagues had spotted sulfur. Showing the graphic trace of the spectrum, Yelle pointed out its larger ripples. The fact that the ripples were spaced regularly indicated a simple molecule. Next, the ripples were close together, the signature of a heavy one. "At 3:00 this morning, we zeroed in on sulfur," he concluded. This is significant because it probably confirms the existence of a long-suspected layer of ammonium hydrosulfide beneath the upper ammonia cloud layer. Also, sulfur, in the form of hydrogen sulfide, exists deep in the atmosphere.[5] Although there was still some question as to whether the sulfur came from the comet or from Jupiter (sulfur was detected in Comet IRAS–Araki–Alcock in 1983), the scientists didn't think that the sulfur came from the comet.

The series of impacts was taking everyone by storm. From McDonald Observatory, astronomer Anita Cochran reported spec-

*The comet's three discoverers discuss a point during one of NASA's Daily Comet Updates.
Photo courtesy NASA.*

tacular skies and stable observing conditions—unusual for south-west Texas in summer. "We've been running around like giddy kids!" she exclaimed.[6] No one had anticipated what we were seeing. At Lick Observatory in Northern California, Mike Liu was having more interesting experiences. "During the last night," he wrote, "two cops on the mountain came into the control room. They were very curious about all the happenings, wanting to look at the screen, and asked a lot of questions about Jupiter."[7]

A different group of astronomers was combining the old and the new. They had successfully mated the old 36-inch-diameter Crossley telescope on California's Mount Hamilton with a high-speed CCD system. One year shy of its centennial, this grand telescope saw its first light in 1885, so it was probably the oldest telescope being used for professional observations in the S–L 9 campaign.[8]

Maybe for professional observations it was, but the Crossley was certainly not the oldest telescope being used just to look at Jupiter that week. In Washington, many people were using the U.S. Naval Observatory's 1873-vintage 26-inch reflector.

JULY 19: FRAGMENT L COLLIDES WITH JUPITER

On the afternoon of July 19, just after 3:00, I began talking to a group of teachers at the University of Maryland. Like educators everywhere, they were always searching for ways to make their subjects come alive, so they shared their exhilaration, feeling that something tremendous was happening that their students would enjoy come fall. But at 4 p.m. I put my lecture on pause: It was time to note, I said, that at that very moment, fragment L was hurtling at 135,000 miles per hour, only two hours away from its destruction in Jupiter's atmosphere. Eventually, Whately Observatory detected the infrared flash. "In addition," they wrote a couple of hours later, "we looked through the 4" refractor which serves as the finder for our 16" telescope. The L impact site, which is now at the center of the disk, appears as a distinct dark spot. We also believe we can see site G as a slightly more diffuse spot to the east."[9]

L's swan song was generating considerable excitement around the world. A multinational team of French, Spanish, and Swedish observers at the Nordic Optical Telescope in the Canary Islands reported that L might have made an even bigger splash than H had.[10] Although their telescope was only 16 inches (0.4 meters) in diameter, L's plume saturated their detectors. They switched to a denser filter, also centered around 2.29 microns in the infrared, but the plume continued to brighten until the image saturated again, despite that filter's greater screening ability. Several more minutes passed before their detector could accept all the light that the plume was throwing at it.[11]

There was no doubt that the public fervor was mounting as each event played out. Planetarium executives returning from their convention in Florida found a huge demand for impromptu

What if fragment G hit Earth instead of Jupiter, say at the north pole. This illustration shows the G impact site taken with the Hubble Space Telescope, but superimposed on Earth. The expanding ring is rapidly crossing the planet, and below the ring is the dark crescent which already extends well into Earth's southern hemisphere into the south Atlantic and Africa. Just two hours after impact, the Earth's upper atmosphere would be filling up with impact debris. Image courtesy John Spencer.

public skywatching parties. "I've heard many people say that Jupiter has become much more of a world to them," said Ian McGregor of Toronto's McLaughlin Planetarium.[12]

Around the world, public facilities faced turnouts numbering in the thousands. Astronomer Mark Sykes was amazed at the size of these crowds in Tucson and wrote, "As a community we need to take full advantage of this remarkable event and its visibility and encourage the public to look at it through their own telescopes (of sufficient aperture) or go down to their local planetarium or public observatory to look at it. It is not often that we have an opportunity to share with the public such an exciting event in a very direct manner."[13]

As other astronomers responded to Sykes' plea, it was clear that the astronomical community had shouldered the responsibility that busy week of keeping the public informed of what was going on. "I want to second Mark Sykes' call for encouragement of public involvement," wrote Tim Livengood from "a very cloudy and grey European Southern Observatory in La Silla, Chile.... We must also encourage the scientific value of amateur astrophotography. Amateurs will be able to compile a much more continuous record than large telescopes can manage over months or years."[14] Amateurs have that advantage because they have access to their telescopes over long periods of time.

Observers from both famous Cambridges reported how last-minute skywatching parties had attracted hundreds of people who wanted a look at the new Jupiter. In Massachusetts, MIT asteroid scientist Rick Binzel reported that an extemporaneous star party on the MIT campus—held with virtually no advance publicity—had brought out 300 viewers who saw very dark markings from spots G and L, the most dramatic impact sites, through small telescopes. "Many of us are finding ourselves advising the public on what may or may not be visible through small telescopes and are working to organize star parties," he wrote. "It would be helpful for the public effort if [observers] will also continue to describe the relative visibility of impact features."[15] From Cambridge, England, John Rogers agreed: "Mark Sykes will be pleased to know the 'public involvement' was heavy; we were glad to show the scars on Jupiter to lots of visitors, though it did compete with serious observations!"[16] The telescope could be switched from scientific observing to public viewing by inserting an eyepiece. Undoubtedly all this interest was a direct result of the impact-to-impact media coverage. At morning and evening newscast time, one could tune in virtually any television or radio station and catch a scene from the latest press conference.

While most observers were absorbed in watching the present, David Jewitt and Paul Kalas were peering into the future from

14,000 feet up on Hawaii's Mauna Kea. Using a red filter and a coronagraph, they succeeded in getting images of some remaining fragments of S–L 9 as they passed inside Jupiter's magnetosphere before their impacts. Surprisingly they found tiny P2, both of the Q fragments, R, S, and the W caboose. But the fragments had changed. Instead of the dust tails of previous months, the fragments now looked like tadpoles, with their little tails pointing toward Jupiter along the axis of the comet string. The remaining comet fragments, Jewitt pointed out, seemed headed tail-first into Jupiter.[17]

At almost the same time, John Rogers chimed in with observations about the spots left over from the earlier hits. Again a surprise: Two days earlier, the Hubble had seen spots E, A, and C almost evenly spaced. But now it appeared that E was wandering around slightly. Possibly Jupiter's upper atmosphere winds were affecting it more than they had affected the other spots.[18]

At the end of this day, a first report came in from the team monitoring Voyager 2, now far out in space but still looking back toward Jupiter with its ultraviolet spectrometer. Possibly just because it was so far away, the spacecraft didn't see any significant changes on Jupiter from any of the several impacts it observed.[19]

JULY 20: THE TRIPLE WHAMMY!

Ten hours before the components of Q slammed into Jupiter, the Hubble Space Telescope captured them as images. It was an eerie sight. To those of us given to anthropomorphizing comets, the two fragments appeared to be embracing Jupiter; their comae stretching out in both directions along their paths to the planet. Hal Weaver again made the point that the Hubble images showed no evidence that the comet nuclei were breaking up, only that their comae of dust were elongating. Their assault on the planet would begin near noon, eastern time, on July 20. It didn't take much figuring to conclude that since Jupiter's rotation period closely matched the intervals between impacts of Q1, R, and S, these three comets would land virtually on top of each other. But when Heidi Hammel brought up this point at the July 18 press conference, the media latched on to it, and headlines like "Jupiter Faces Triple Whammy!" blazed around the world.

"Q1 is going to go in," Heidi Hammel said, "and exactly one Jovian rotation later—10 hours—R is going to hit right next to the very same longitude, and one Jovian rotation after that, S is going to hit the same longitude. So you're going to have three—BOOM, BOOM, BOOM—right on the same small range of longitudes. And that," she concluded as she punched her fist into her hand, "is going to make one heck of a mess."

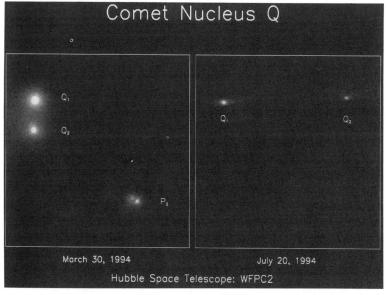

The evolution of fragments Q1 and Q2. On March 30, 1994, the two fragments were bright and circular. On July 20, the two fragments are just a few hours away from their crash into Jupiter. Jupiter's gravity has already started to stretch the comae of dust particles surrounding them, so they appear elongated.

For a planet already scarred with beating after beating, this new triple assault promised even more pyrotechnics. Also, since Q had been by far the brightest fragment before it split into two parts, many of us had expected that it would pack the biggest punch.

First came tiny P2, forming a small spot virtually atop the enormous K site. Q1 was bright, but not as bright as L, reported the team of French, Spanish, and Swedish observers at La Palma.[20] From Calar Alto, observers detected two spotlike features, the second one (Q1) being much brighter—in infrared light—than the first.[21]

Then came a reliable report from the Keck telescope atop Hawaii's Mauna Kea, where observers had seen the impact from fragment M. This was a startling observation. Since M had not appeared on any telescope images since January, virtually everyone thought it had completely broken apart. However, a few minutes after M should have hit, had it existed, the Keck team detected a new series of very small spots.[22] It appeared that even though the fragment had faded too much for even the Hubble Space Telescope to see, it had still been there, and it had still crashed into Jupiter. Gene and Hal Weaver rejoiced at a later press conference, because this confirmed the stand they'd taken: When a comet disappears from sight, has it really dissipated into nothing, or does it simply turn off? Fragment M's revived presence on Jupiter demonstrated that at least this vanishing comet simply had become too faint to see.

Although Q1 did not leave as large a spot as the giants G, H, K, and L had, it and its neighbor Q2 left a complex set of markings. The triple whammy had become a quadruple whammy. One of the new spots seemed to be the remains of Q2, but the Hubble science team couldn't be a hundred percent certain of that. "That is either Q2," said New Mexico State's Reta Beebe the next day, "or it is a Q-lette!" It was the smallest impact site observers had seen, even smaller than D was when it was first formed. But the interesting thing was that even though it was small, it was behaving in shape and rate of expansion in much the same way as the other spots.[23]

Unfortunately the weather in Chile suddenly turned bad, and the great telescopes at the Cerro Tololo Interamerican Observatory and at the European Southern Observatory were completely out of business.

SHOEMAKER–LEVY 9 GOES TO THE WHITE HOUSE

For a brief moment during the events of July 20 my mind went back to a different place and time. On July 20, 1969, I was in an auditorium with more than 100 children at Camp Minnowbrook on Lake Placid, New York, all watching a tiny 12-inch black-and-white TV mounted on the stage. For several hours, even the youngest children sat transfixed as Neil Armstrong stepped down the ladder onto the dusty surface of a world so ancient, and yet new to us.

Twenty-five years later I sat in another auditorium, this one at NASA's Goddard Space Flight Center. I was part of the daily NASA comet update for the media, which by now was attracting an ever increasing audience share as more and more people focused their attention on Jupiter's time of trial. During this press conference I strongly encouraged everyone with a telescope to go out to look at Jupiter, a planet now girdled with impact spots. "No matter where you are on Earth," I said, "if you look at Jupiter in the evening sky you should see some spots."[24] In the back of the room, Gene and Carolyn had arrived from a meeting about the congressional call for a search for asteroids that could endanger the Earth. Gene was beckoning to me, but nothing could make me leave the podium until after the press conference.

Finally, while someone on the other side of the panel was talking, Gene approached me. "We gotta get out of here, old boy!" he whispered. "Why?" "We're going to the White House! Goddard has a car to take us there!" Well, nothing would make me leave—except that. As we walked out of the auditorium I gave Mother a hug. "Have fun!" she smiled. I wish she had been invited too.

The White House had planned a ceremony to note the 25th

anniversary of the landing on the Moon, and the discoverers of Shoemaker–Levy 9 were invited to attend. From the moment we walked into the national landmark we were extremely well treated, and as we met members of the staffs of both the President and the Vice President, we felt very much the center of attention, even though the ceremony was not supposed to be for our doomed comet.

The East Room was elegantly arranged with Apollo decorations and a beautiful wall-sized photo of Buzz Aldrin and the American flag on the Moon. For Gene, the celebration had a special poignancy. Except for Harrison Schmidt, who was already a professional geologist when he joined the program, the Apollo astronauts owed their geologic training to Gene Shoemaker. He had taken each of them, including today's honorees, Neil Armstrong, Buzz Aldrin, and Michael Collins, to Meteor Crater and to other moonlike places. Gene thought that Armstrong's 90-minute walk on the moon was, despite its brevity, one of Apollo's most productive triumphs. "He made more pertinent observations in the precious little time he had on the surface," Gene recalled, "than many of the astronauts who followed him."[25] During the White House ceremony, it was obvious that Gene's high regard for Armstrong was reciprocated, as the reclusive astronaut clearly returned Gene's good feelings. "Our old astrogeology mentor, Gene Shoemaker, even called in one of his comets to mark the occasion with spectacular Jovian fireworks," Armstrong said.[26] President Clinton enjoyed his tribute.

After the ceremony, a White House staffer approached us. "The President wishes to congratulate you," she said. "Please wait here." I turned to Carolyn: "When you first found our squashed comet, did you expect that it would lead to this?"

"Well, of course!" Carolyn bantered. "Didn't I say 'David, this looks like a squashed comet. Call the White House and tell the President?' " A few minutes later President Clinton approached us, smiled, and congratulated us on our discovery. After he left, the room emptied quickly. We were about to leave when another staffer walked up. "The Vice President wishes to speak with you," he said. "Would you mind waiting a few minutes while he finishes a radio interview?" While we stood by, we learned something few people were aware of: Al Gore had a lively interest in reducing the amount of unwanted light that is a form of pollution over our cities. Light pollution is a serious threat to the environment; each year the United States spends some two billion dollars to light the underbellies of birds and airplanes. Besides being wasteful, unwanted light does affect the darkness of the night sky. These facts are not widely known outside the astronomical community. I was impressed that Gore would have such a keen awareness that the darkness of the night sky is a delicate part of our natural environment.

Having completed his radio interview, Gore approached us with a wide smile. "I really want to meet people as famous as you three are this week!" We had an animated conversation, and I was delighted with the Vice President's keen interest in the impacts. He had obviously been keeping up with what was going on. His personality was so disarming that I had to remind myself that this perceptive amateur astronomer—one of many this week—was the Vice President of the United States.

This visit to the White House to meet the President and the Vice President was a highlight in a week of highlights. Our return trip was not quite as comfortable as our ride to the mansion. In fact we had to wait outside for a cab in searing heat and humidity.

We caught our breath at the hotel, but soon it was time to dress for the Planetary Society gala dinner. Black tie is not something I am comfortable with on a hot July evening, but with everything else happening this week, a night in a monkey suit wasn't a problem. Since Gene and Carolyn were staying in Baltimore, not Washington, my mother and I offered them our rooms to change. Gene came in to my room carrying his rented tux just as I was returning a call from a journalist from the Phoenix Gazette. As I answered the reporter's questions, I heard Gene muttering. I turned around and saw him trying to get into a jacket with the tailor's pins still in place. We looked at each other in some surprise, and then I went back to my interview. A minute later I turned around again and saw Gene almost in shock—his trousers stretched at least eight inches past the tips of his shoes.

Seeing the man of the week—praised within the hour by the first human to step on the Moon and by the President of the United States—stand in this sorry suit of clothes was too much after all the tension of the day. And trying to stifle a laugh made it worse. I couldn't finish the interview and had to tell the reporter I'd call her back. I really did need to help Gene out of this unseemly situation. Soon Mother, Carolyn, four friends, and the hotel's seamstress were all on the scene. "Dr. Shoemaker," the valet said as she looked at his casual brown walking shoes, "I would be able to measure your trousers better if you put on your dress shoes."

An exasperated Gene looked down. "These are my dress shoes," he announced.

Within another hour Gene was properly hemmed, fitted with borrowed black dress shoes, and ready for the Planetary Society dinner. I even got to finish my interview.

With the Vice President's speech as a highlight, that evening was quite a success. Gore stressed how many of NASA's diverse space missions and ground-based activities had came together for Shoemaker–Levy 9, a point that echoed and expanded on what I had been saying in interviews all week. Gore's comments reflected another idea that I share—that NASA should expand its role in the education of future scientists and, by inference, that this week's impacts were an ideal launching pad for this lofty enterprise.

Gore's speech was followed by a revue from the Capitol Steps, a famous satirical troupe making fun of political events. The Shoemakers and I enjoyed the show, but there was no rest for us. We slipped out and headed to the hotel's second floor to take part in PBS's hour-long special on Shoemaker–Levy 9. For me at least, the program was controversial. During the entire hour I apparently failed to mention the name of the Association of Lunar and Planetary Observers, an amateur organization whose members had done yeoman service observing these impacts. Even though many amateur groups had involved in the effort, one ALPO member went on and on about my leaving them out. It was a small criticism, but by this point in the most demanding week of my life, my emotions were getting a bit raw—from lack of sleep, laughter over Gene's tuxedo, to excess worry over how I'd handled a reporter's question. Notwithstanding the many high points, this was the most difficult week of my life.

As the day ended, I finally went outside to see that the sky was clear, and that once again I had not had a chance to look at Jupiter. But not everyone in that banquet room was as unlucky as I. After finishing his speech, Vice President Gore rushed to his home at the Naval Observatory and peered through a telescope at the spots on Jupiter.

13

Second Star to the Right, and Straight on Till Morning

The magic of impact week was that it was a time for elated astronomers to act like children and for children, who turned their new telescopes on Jupiter, to act like scientists. By Thursday, July 21, all but a handful of comet fragments had reached Neverland on Jupiter.[1] The planet's southern hemisphere was getting so crowded with the remains of the big fragments G, H, K, and L that anyone with almost any telescope could see them. The comet's bravura performance had captured the imagination of our world, and as S–L 9 prepared for its grand finale, newspaper editorials and cartoons noticed the magic of its troubled course. "Jupiter and Shoemaker–Levy 9 made cosmologists of us all," wrote Michael Skube in the Atlanta Constitution, "witnesses to the greatest show not on Earth."[2]

BATTLE DAMAGE ASSESSMENT

Back at the Naval Observatory, Jim DeYoung was using the 24-inch reflector to observe S–L 9 and report accurate positions of the comet in space. "Since I became the local 'expert' I was requested to give a briefing to an Office of Naval Research group. I arrived with lots of nice slides to describe what was going on, the physics, etc. I was introduced as, '…Jim DeYoung from the U.S. Naval Observatory who will be giving a BDA on the Comet Shoemaker–Levy 9 impacts.' Not knowing what a BDA was, I had to ask. I was told a BDA is a battle-damage-assessment!"[3]

At Williams Bay in Wisconsin, Walter Wild had planned an ambitious program using Yerkes Observatory's 41-inch telescope. One of the oldest observatories in the United States, Yerkes was built by George Ellery Hale—the famous astronomer and fundraiser for telescopes. Hale built the world's biggest telescope over and over again—first it was the 40-inch at Yerkes, which remains the biggest operating refractor on Earth. Most big telescopes are reflectors, using mirrors instead of lenses. Then it was the 100-inch Hooker reflector at Mount Wilson, and finally the 200-inch at Palomar. One of the telescopes at Yerkes was a 41-inch reflector, and Hale would have been amazed at how the observing team was planning to use it. Coupled to the telescope was a system of optics that would actually be intelligent. With this system, the telescope mirror actually would change its shape in response to the steadiness of the atmosphere. They made a heroic effort to have the instrument aligned and operating by July 16, but since testing was delayed by a spell of bad weather, their first nights were not very productive.[4]

More than two thousand miles away, in Las Cruces, New Mexico, Clyde Tombaugh was also looking. Tombaugh had observed Jupiter since 1918, so he was pretty familiar with the giant planet's features. In 1930, his assiduous searching added the new planet Pluto to our roster of planets in the solar system. In the postwar years he used his knowledge of telescopes to develop the optical tracking system for the missile program at White Sands Missile Range. "I expected a big rumpus from these impacts," he recalled. "I have watched missiles hitting the ground at velocities around a mile per second. The energy displayed was awesome—the impacts made craters 75 feet wide and 20 feet deep, and splashed out debris for a quarter of a mile." To observe Jupiter that night, the 88-year-old Tombaugh used his portable 10-inch reflector. To make this telescope easy to move around, he ingeniously mounted it on a lawnmower chassis. On this night, Tombaugh's first look confirmed his suspicions. "What surprised me was how dark the spots were. I had expected they would be hot, and white."

At the big observatories in Chile, clouds and ice prevented any observing. Thousands of miles northwest, at 14,000 feet atop Hawaii's Mauna Kea, the night of July 21 was cloudy and windy. Churning offshore was Hurricane Emilia, and all observatory domes were being evacuated this night. Steve Edberg had flown there, hoping, he wrote, "to join Lonne Lane's merry band and assist with observations with the Ultraviolet Spectrometer/Imager mounted to his portable 16" Cassegrainian telescope that could be wheeled in and out of the dome of the University of Hawaii's 88" telescope. My hope was to observe W firsthand, since it would be so close to the limb and Jupiter would be above the horizon in Hawaii.

"Alas, this was not to be. With a hurricane blowing by the island, a tour of the (closed) Keck 10 m telescope observatory was all I managed that night, with some of their earlier impact image sets played on a monitor. Only on the last of my four nights there did we get to observe."[5]

WATER

On two days during July, Ann Sprague, Donald Hunten, and their colleagues were aboard the Kuiper Airborne Observatory as it lumbered down a runway in Melbourne, Australia, to chase fragments R and W. It climbed to 41,000 feet, which is higher than virtually all our atmosphere's carbon dioxide and water. Twelve minutes after R crashed, they detected a spike in their spectra, which they later identified as water vapor, at the high temperature of 500 Kelvin. They detected enough water to fill a 50-meter-wide

The Kuiper Airborne Observatory flew several missions during the impacts of Shoemaker–Levy 9 and contributed vital data, including the discovery of water in several impact plumes. Courtesy A. Sprague and D. Hunten, University of Arizona.

sphere of ice, and a slightly lesser amount of water from the W impact. Whether the water came from the comet or from Jupiter was still uncertain by the end of January 1995.

Although the comet was grabbing the lion's share of the news that week, there was yet another big event going on. The space shuttle Columbia was in orbit around the Earth 13 years after its first launch. The craft carried the International Microgravity Laboratory, complete with jellyfish, goldfish, and even newts from Japan. But in the press, Columbia's various life forms were no match for the fireworks on Jupiter.

---→

Carlos Hernandez, an amateur astronomer from Miami, Florida, sketched Jupiter on each night during the impacts. This series of drawings is his interpretation of the changing picture of the planet during impact week.

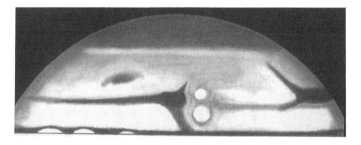

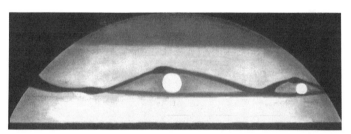

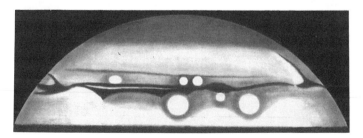

Lucy MacFadden and David Levy discuss the visual appearance of Jupiter at the Daily Comet Update of July 22, 1994. Photo courtesy NASA.

The press was having great fun at Jupiter's expense, in part because of the underlying threat of a similar barrage here on Earth. Besides, where else could one witness such an explosion without accompanying human devastation? Also, the participants at NASA's *Daily Comet Update* were so lively and their explanations so lucid. Heidi Hammel's "BOOM, BOOM, BOOM!" was reported in almost every newspaper in the world, as was Gene Shoemaker's answer to "What is the most important thing we learned?" Looking straight at his audience, Gene quipped, "Yes, Virginia, comets really do hit planets."

"On behalf of 5.5 billion Earthlings," a Maryland newspaper intoned, "we kindly ask all comets out there to please stay over in that end of the solar system. Thanks for your cooperation."[6] Some news writers commented on the excitement. "We don't pretend to have a firm grasp on the scientific implications here," wrote an editor of the Binghamton, New York, *Press*. "But we do know genuine excitement when we encounter it, and the scientific community is genuinely excited … Shoemaker–Levy 9 should help to remind us just how much of our universe is still unknown and inviting to the right combination of imagination and education."[7] In the *Boston Globe*, another editorial carried this message: "Earthlings may content themselves with the wonder of having watched a natural disaster on a scale unparalleled in historic times from the safety of 400 million miles distance—close enough for a good look, far enough to avoid the splash. A pretty good arrangement."[8]

"When the world seems to be spinning out of control," quipped the Primos, Pennsylvania, *Times*, "and the major issue on everybody's mind is O. J.'s innocence or guilt and whether Michael Jackson married Elvis Presley's daughter, it's refreshing to know the heavens have something else to offer … the collision of Comet Shoemaker–Levy 9 with Jupiter."[9] And the *Kansas City Star* noted: "The explosions caused by that comet hitting Jupiter have been spectacular. The surprise is that we haven't seen any 'I survived the Shoemaker–Levy 9 comet' T-shirts."[10] From the *Advance* out of Elizabeth City, North Carolina, came this more elaborate thought: "The events of the past week have helped to reinforce mankind's need to reach into the heavens and explore the unknown.… We should use the inspiration … to reconfirm our commitment to progress and to further our exploration of outer space. It is the human thing to do."[11]

Meanwhile, in Taunton, Massachusetts, a dog attacked a 7-year-old boy. Fortunately, the child's wounds were minor, but the animal control officer at the scene blamed the attack on "a combination of the weather, the Moon, or the Jupiter thing." Just in case the comet proved innocent, however, the officer also promised that he would talk to the owner about the dog's behavior.[12]

Right at the heart of this media hurricane were we three discoverers, and by Thursday, July 21, we were utterly exhausted. The press conference that day was just as assiduously covered as all the others. Public interest seemed to increase rather than wane. Except for the first two *Daily Comet Updates*, which took place on

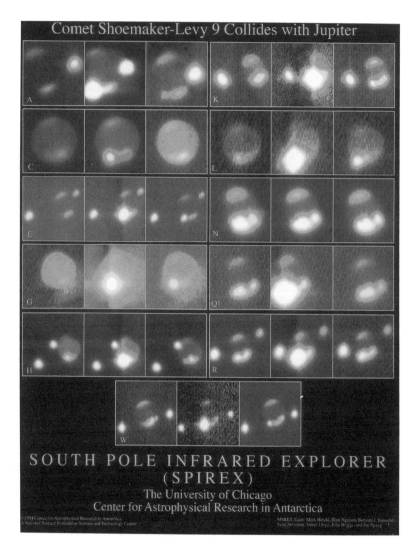

July 16 at the Space Telescope Science Institute, all the conferences occurred at NASA's Goddard Space Flight Center against a backdrop of the January 1994 image of Shoemaker–Levy 9. Each day when we looked at that image, we noted the fragments that had ceased to exist. Next door, my Canadian friend Peter Jedicke, who'd driven down from Ontario to share the fun and help me through the wild seven days, happened to be near the bank of phones set up to handle the flood of media inquiries, when a reporter from Toronto's *Globe and Mail* called with a series of questions. Peter was quickly pushed into service as a knowledgeable Canadian source, and did well providing information and analysis the rest of the week. He helped make sure that the Canadian angle of the story, especially the successful orbit calculation by Ontarian Paul Chodas, was emphasized in Canadian coverage.

Heading right at Jupiter, fragment S was expected to crash just two hours after the end of the press conference. There was a lot to ponder that day, especially a new report from Randall Gladstone of the Extreme Ultraviolet Explorer satellite. The EUVE was investigating the Io torus—a ring of charged particles orbiting Jupiter around the distance of the moon Io. Gladstone reported that at this early date, EUVE had not detected any significant changes in the Io torus.

Some members of the media had been conversant in astronomy before the crash became news. Others were becoming rapidly more learned as the week went on. Recognizing that comets come in different types, they began asking what kind S–L 9 was. Gene explored the model proposed by Eric Asphaug and Willy Benz, which suggested that S–L 9 was a weakly bound collection of

←————————————————————————————————

With the Sun never in the sky and Jupiter always in the sky, the University of Chicago's Center for Astrophysical Research in Antarctica was perfectly set up to follow Jupiter during impact week. Mark Hereld, Hien Nguyen, Bernard J. Rauscher, Scott Severson, James Lloyd, John Briggs, and Joe Spang fought blowing snow and frigid temperatures to capture eleven of the impacts. This series of images shows each of the impacts as a set of three. Usually the middle one shows the impact at its brightest phase. It is interesting that small fragments like N, which was also captured easily by the Galileo spacecraft, produced such dramatic brightenings.

LUMICON NGC NOTES

OBJECT DESCRIPTION

NGC/IC No._____ Other_____

Size_____ Magnitude_____

Object Type ___Jupiter SL 9___

Location ___IMPACT SITES___

EYEPIECE DRAWING

L GSRQ2 Q1 NB H

VIEWING CONDITIONS

Date 7-22-94   Time 3:56UT

Seeing 8   Transparancy 4

Observer's Name Barbara Wilson

Observing Site Home

Telescope Type Newton 10" F-5

Filter Type NONE

Eyepiece 10.5   Magnification_____
plossl 2X barlow

ADDITIONAL SKETCHES & NOTES
(use backside of card for additional space)

System II Long
321.16°

1984 LUMICON

Barbara Wilson's sketch of Jupiter's impact sites typifies what could be seen through amateur telescopes. Wilson used a 10-inch (25-centimeter) diameter reflector on the evening of July 22.

smaller bodies. Gene maintained that it was a rubble pile that had repeatedly smashed up in the past and then reaggregated.[13] Other models would be debated long after this day's press conference had wrapped. One suggested that the fragments were large solid pieces, and another proposed that they were small solid pieces.

This press conference offered New Mexico State University's Rete Beebe a public opportunity to discuss what she had hoped to accomplish. Weeks before the impacts began, Beebe had written eloquently about what she had hoped to see: "If a local region were heated to 15,000 degrees K (27,000 degrees F), it would radiate heat away 100 million times faster than the cold cloud deck and generate an impressive flash. It would cool rapidly in the early stages; but as the spot cooled, it would lose energy more and more slowly. If minimal mixing with its local surroundings occurred, it could be weeks or months before the spot totally lost its identity. Such a hot spot could act as an obstacle to local wind flow. Because the atmosphere of Jupiter is at a temperature where ammonia readily freezes or melts, the induced perturbations should generate cloud patterns that will serve as markers to indicate how the atmosphere is responding to the remaining hot spots. It is this atmospheric response that I have been studying. This requires observations before, during, and after the event."[14]

With the benefit of hindsight on the day of this press appearance, Beebe could see no interaction between the spots and the material underneath. "I wanted to see a bubble come out," she explained. "Then I would understand how much energy it took to run the storms I've been watching for the last 25 years. The brighter material would flow east along the north side of the spot and west along the south side of the spot, as a banded streak. We don't see that."[15]

Suddenly there was a loud crashing sound in the back of the room. We were all quite startled, but no one seemed to be hurt. "Is everyone okay?" someone asked. "Was that S?"

France Cordova, NASA's Chief Scientist, finished the press conference with a special comment: "I'd like to congratulate all the participants in this event, for the inspiration that they continue to give.... Astronomy is about transforming images, those images that turn upside down our world view and that give us a new perspective on ourselves and on nature." Her words articulated what so many of us had been thinking—that more important than anything else, this week was inspiring. "I thank you all for your affirmation that science is fun," she concluded, "if not quite as simple as A-B-G."[16]

After the press conference, Gene and Carolyn went to Washington to be interviewed on a radio show; by then I was so overloaded that I didn't catch which one. As I went out to lunch with a group of Goddard scientists, I was delighted that for one

An exhausted group of scientists at the Daily Comet Update, 23 July 1994, at Goddard Space Flight Center. Left to right: Don Savage, Heidi Hammel, Melissa McGrath, Gene Shoemaker, Carolyn Shoemaker, David Levy, and Lucy McFadden. Photo by Peter Jedicke.

precious hour, I would be out of reach of anyone. We sat down at a Chinese restaurant and started to enjoy our lunch. The waitress took our orders. After a few minutes she returned. "Is there a David Levy here?" I admitted that there was. "You have a phone call," she said.

At the other end of the line was George Cruys. "How did you know I was here?" I asked incredulously. He said that he had overheard our group decide to have lunch at a Chinese restaurant, and it had taken a little detective work to find one near the space center. "You've got to reach Gene and Carolyn," he said. "ABC News has selected the three of you to be featured on 'Person of the Week.'" So much for our quiet little lunch. Finding out where Gene and Carolyn were was easy: We surfed along the radio dial and quickly found Gene's deep voice on National Public Radio's *Talk of the Nation*.

Our moderately busy afternoon became a hectic one. ABC News interviewed me at NASA headquarters, where I was setting up to do my second set of a long series of local television interviews, sent to a dozen stations via satellite. Over the next three hours my back-to-back interviews reached an estimated total audience of 6.9 million viewers. It was challenging to answer a similar series of questions repeatedly while trying to keep the answers fresh and directed toward each particular city. Peter Jedicke, who was watching from the control room, noted that after a broadcast the director would tell me I was off the air, and I would slump down in my chair, thoroughly exhausted. A few minutes later he would say, "Okay, David, you're on the bird!" and I'd snap to enthusiastic attention for the next interview. I enjoyed the afternoon immensely, and I understand that these interviews were NASA's first attempt to conduct such a program. It worked extraordinarily well. I had the feeling that for each 10-minute period I was in touch with a particular local station and was able to aim my comments toward that city's audience.

AN EVENING WITH HISTORY

When my last NASA interview was finished, we rushed across town to the U.S. Naval Observatory for yet another press conference and, at long last, the chance to see the now famous impact spots. We drove away from NASA headquarters under a sky dark with clouds and rain. Would the sky clear before Jupiter set? The window of opportunity, before Jupiter sank too low in the west, would last about two hours.

On this of all nights, it seemed appropriate to observe Jupiter, especially at such a historic observatory. The observatory's long history grew out of the Navy's oldest scientific institution, the Depot of Charts and Instruments, which dates to 1830. Completed in 1844 at Foggy Bottom (where the State Department now is, in Washington) the observatory moved to its present site in 1893, and last year it celebrated its centennial at that site.

In the summer of 1862, a young astronomer named Asaph Hall was appointed an "aide" at the Naval Observatory. Hall was on duty at the 9.6-inch refractor telescope at the old Foggy Bottom site. Soon his work was interrupted by the explosions of the second battle of Bull Run, and he helped care for friends wounded there. He was "observing Mars every other night," his son wrote, "and serving Mars the rest of the time!"[17]

By May 1863, Hall was promoted to the status of professor, and his work schedule settled down. As he began observing on the clear night of August 22, a knock at the trap door that separates the observing floor from the anteroom below startled him. The tall visitor was Abraham Lincoln. Depending on which source we use for the tale, he was accompanied either by his personal secretary, John Hay, or his Secretary of War, Edwin Stanton. During the brief observing session that followed, Lincoln got a good look at the Moon and, for good measure, the bright star Arcturus.

There is more to this story. Hall was observing at the 9.6-inch telescope a few nights later, when he heard another knock at the door. Not expecting anyone in particular, Hall finished his leisurely observation before lifting the door. Poking his head through was the President of the United States, this time completely on his own. He had walked through poorly lit Washington from the White House. Apparently it was his habit to walk alone occasionally at night. Passersby would doubtless tip their hats and say "Good evening, Mr. President."[18]

Lincoln had returned because he was puzzled. When he used an old surveyor's telescope, it showed the Moon right side up, the way it was supposed to look. Why did the refractor show it upside down and reversed? It would have been fun to listen to Asaph Hall answer the President's question, which is still the single most asked query about an astronomical telescope. The surveyor's telescope, Hall would have explained, had an extra lens to correct the image. Astronomical telescopes do not use such lenses.

In 1877, Asaph Hall had used the 26-inch to discover Mars' two tiny moons, Phobos and Deimos. It was in this magical environment that we hoped to observe the impact sites on Jupiter.

Our weather in Washington, though cloudy and raining, was hardly as dramatic. But unlike the astronomers at the other observatories, we did need Secret Service clearance to get to our telescope, for we were at the U.S. Naval Observatory. Since Nelson Rockefeller chose it for his residence during the Ford Administration, this special observatory has been the home of the Vice President of the United States, who resides in the mansion once used by the Admiral of the fleet.

I thought that on this night of July 21, 1994, if only the sky would clear, hundreds of people might see the most incredible view of Jupiter ever had with the Naval Observatory's historic 26-inch telescope. Also, a squadron of smaller telescopes decorated the observatory's front lawn. Bob Summerfield and Lisa Levicoff ran the largest of the amateur telescopes as part of their Philadelphia operation called "Astronomy to Go." Their group brought some substantial telescopes. "Sharing the sky with the public is what my whole life is all about," Summerfield declared passionately.

As the crowd gathered, the thick clouds resolutely proscribed any view of stars. After yet another press conference, I had just begun a lecture about the impacts when someone alerted me to the clouds breaking up. With one quick sentence I ended my remarks and we rushed up the stairs to one of the historic telescopes. Though I took my place at the end of the line, I didn't argue much when Lou Friedman, executive director of the Planetary Society, offered me the first look.

A chill went through me as I peered through the great old telescope and saw the new Jupiter. Out, damned spots! I could not believe how obvious they were. I quickly released the telescope to someone else and looked through the small finderscope. The spots were just as obvious with it. Wanting to sketch their shapes and locations, I remembered that I had left my observing log downstairs where I had given my lecture. Lou offered to run back and retrieve it for me. When he returned I began to draw the planet. I was overwhelmed by the density of the L and G sites, which were just moving onto the side of Jupiter now visible.

Evolution of D/G Comet Impact Sites on Jupiter

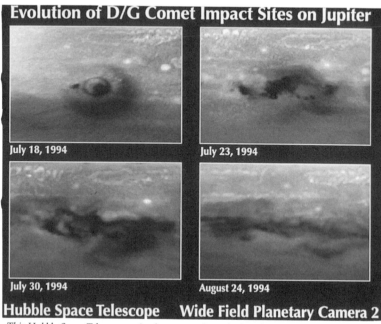

July 18, 1994   July 23, 1994

July 30, 1994   August 24, 1994

Hubble Space Telescope   Wide Field Planetary Camera 2

This Hubble Space Telescope series documents the evolution of the G impact site as strong winds affected the size of the cloud of debris. In the first image, the small dark spot to the upper left is the D site; the D fragment hit at almost the same place two Jovian days (or about 20 hours) earlier. Courtesy Hubble Space Telescope Comet Team and NASA.

Despite the grandeur of the old telescope, the people who peered through the portable telescopes out on the observatory lawn seemed to be having even more fun. One seven-year-old boy approached me smiling from ear to ear. "I just saw three great big spots on Jupiter!" he declared.

"Where were they on Jupiter?" I inquired.

"Right along the top! I saw the comet crash spots!" As he went down the line, telling people what they would see through the telescope, I thought, score one for the comet. "The ease with which the new dark spots on Jupiter can be seen by anyone with a telescope over 2 inches in diameter," one astronomer said, "seems hardly newsworthy anymore!"[19] Thanks to these spots, many young people would enjoy astronomy for the rest of their lives.

FRAGMENT R HITS CLOSE TO JUPITER'S LIMB

Months before the impacts, Paul Chodas and Don Yeomans proposed that the later members of the comet train would be hitting Jupiter progressively closer to its limb than the earlier ones. So it was not too much of a surprise when three observatory teams independently detected what they interpreted as the meteor flash from R's impact. At 5:33 UT on July 21, Imke de Pater and her team at Hawaii's Keck Observatory "detected the first FLASH from impact R: it lasted for maybe 20 seconds: truly remarkable."[20] Some 45 seconds later they detected a second flash. Approximately 8 minutes later the plume appeared, brightened considerably, and faded again.

Around the world in Australia, the observing team at the 3.9-meter Anglo-Australian Telescope caught the R fireball *in daylight.* "A distinct, bright, point-like source first appeared on the morning limb at 05:34 UT. This feature was tentatively identified as the impact fireball from fragment R. At 05:42, as the seeing improved, this feature brightened dramatically, saturating the detector at 05:45:33, and producing distinct diffraction spikes, like the fireballs associated with the impacts of fragments G and K observed earlier this week."[21]

At Palomar Observatory, a team using the 200-inch Hale Telescope observed fragment R's hit. From the southern California site, Jupiter was getting low when a faint brightening appeared at 10:35 Pacific time, or 5:35 UT.[22] That was two minutes later than the observation at Keck. Out in west Texas, McDonald Observatory reported a flash at 5:41,[23] the same instant that Palomar recorded a rapid brightening that continued for six more minutes. Over the following five minutes the impact site continued to brighten until it saturated the telescope's detector, producing beautiful cross-shaped diffraction spikes. After a few more minutes the plume

Almost an hour and a half after fragment R hit Jupiter, astronomers using the University of Hawaii's 2.2-meter telescope on Mauna Kea recorded this image. Jupiter's moon Io is at upper left. From left, the impact sites shown are H, Q1, R, G (with D) and L. Courtesy K. Hodapp, J. Hora, K. Jim, and D. Jewitt; processing by L. Cowie and R. Wainscoat.

started to stretch out along Jupiter's limb.[24] Was it possible that fragment R had broken into at least two pieces months before it plunged to its death? Apparently, at least two teams recorded actual meteors brightening—in the infrared—as they soared through Jupiter's upper atmosphere. At 5:42 UT Marcia Rieke and a team from Steward Observatory's 90-inch telescope atop Arizona's Kitt Peak saw the first indications of the plume.

EXPLORATION BY SPACECRAFT

From their unique viewpoints beyond the Earth's atmosphere, spacecraft beamed back some of the most interesting information about the impacts. Launched in 1978, the International Ultraviolet Explorer, "the little spacecraft that could," was still operating and returning observational data during the summer of 1994, many years after its useful life had been expected to end. Jupiter was one of the spacecraft's regular observation targets. In early June the intrepid IUE began watching Jupiter intermittently; since July 16 the spacecraft had been monitoring Jupiter 24 hours a day. The craft recorded a tremendous amount of energy as the fragments traveled through Jupiter's upper atmosphere.[25] The increase in auroral activity IUE detected could be associated with either direct impacts or the passage of comet material through Jupiter's magnetosphere.

The spacecraft's telescope performed observations that were unique to the entire observing campaign. By recording the emission of energy in the ultraviolet off the edge of Jupiter and along the trail of the comet, the spacecraft was also doing other ultraviolet experiments. According to scientist Walt Harris, the impacts were giving the IUE team "new insight into the characteristics of features long studied with the IUE, and the operators of the other instruments an opportunity to compare the results of their observations with the more than 15 years of Jovian FUV [far ultraviolet] spectra in the archives."[26] The spacecraft continued regular observations of Jupiter until the middle of August, when Jupiter be-

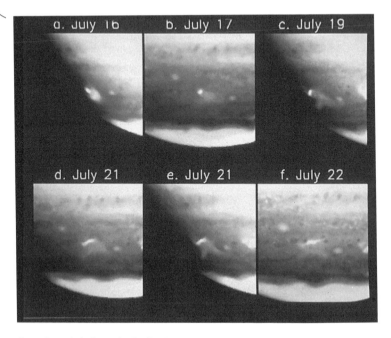

Over the period of a week, the first impact site evolves rapidly as it is buffeted by Jovian winds. This series of images the A site was taken by the Hubble Space Telescope. Image courtesy of Hubble Space Telescope Comet Team and NASA.

came too close to the Sun's position in the sky for safe observation with the satellite.[27]

At the Space Telescope Science Institute, astronomers were now identifying the compound carbon disulfide in their spectra, beyond their earlier finds of sulfur and hydrogen sulfide. Scientists had hoped to find this material, which is supposed to form when hydrogen sulfide reacts with methane during a high-temperature and high-pressure event like an impact.[28] Further, the detection of carbon monoxide by a number of observatories indicated that temperatures in the impact areas had reached several thousand degrees.[29]

From its unique observation post high above the Sun's south polar region, the Ulysses spacecraft also had a direct view of the impact areas. However, it did not detect any obvious decametric radio emission from the crash sites.[30]

BACK ON EARTH

Calar Alto came in with an early report of the crash of fragment S, citing a small plume that quickly became brighter than the nearby spots from earlier impacts, but which faded after a few minutes.[31] Fragments T, U, and V were even less showy, revealing neither plumes nor spots. But considering everything that had come before, the weak performance of the end of the train was a disappointment only because V was one of the very few impact events that took place while Jupiter was in the evening sky of the Americas. "Fragment V did not show any obvious flash or plume in McDonald observations. This makes two fragments in a row for which nothing was detected," wrote observers from Texas, "despite continuous monitoring around impact times, photometric skies and good seeing."[32]

But this was not a time to let down our guard. "Although the U and V fragments appear to have been fizzles, as was B, note that W may not be," wrote the Jet Propulsion Lab's Paul Weissman. "Remember that A was much brighter than expected. W, being the last in the line, may contain an excess of material from near the surface of the former comet nucleus, which was devolatilized and crusted over before the breakup, resulting in a fragment that was incapable of being as active as the other ones from deeper inside the nucleus. [If W had a thicker crust than the others, Weissman maintained, it might shine less brightly. Therefore it could have been a bigger fragment than it appeared to be.] If W is large and inactive, it will still create quite a blast when it enters."[33] While others felt that W's crust, if there was one, was only a few meters thick and should not have affected the activity of these outer particles, the idea was a possible explanation of why A, which went down what now seemed so long ago, was so dramatic.[34]

To conserve its limited data storage capability, the Galileo Spacecraft solid-state imaging program took this long series of Jupiter images that appear as a time lapse in several rows. The brightening of the plume from the impact of K is seen as a thin line to the left of the planet starting in the third row.

"It's just about an hour to impact here," Weissman concluded. Good luck on the final shot." Since Weissman was a scientist on the Galileo mission, and the distant spacecraft was at that moment pointing its camera toward Jupiter, he had good reason to be alert for W. The plan was to take a series of snapshot pictures of the final impact that might reveal the flash of the meteor as it plummeted into Jupiter's atmosphere. As fragment W approached Jupiter, the spacecraft began a series of exposures.

W was coming in fast. For a worldwide team of exhausted astronomers, there was just this last explosion to catch, one more set of events to record. Just one more, but W was very significant. According to Don Yeomans and Paul Chodas, W offered the best

Night of the Comets. Several smaller fragments, possibly pieces of Fragment P_2, enter the sky near Jovian dawn, the fireballs illuminating the dark clouds. "Another lights up a nearby thunderhead," the artist writes, "like a Chinese lantern." Painting by Kim Poor.

hope of showing something visible to Earth as it flew through Jupiter's atmosphere.

It was almost precisely 4 a.m. on the East Coast of North America, where astronomers were working with the Hubble Space Telescope; 1 a.m. on the West Coast, where astronomers working with the Galileo spacecraft had little to do but wait. Observing commands for both crafts had been uploaded weeks before. The Hubble data would be available within hours, but the results from Galileo, through its snail-paced antenna, would not come down for weeks. The two spacecraft were programmed independently. No one could have predicted that at one of the most critical times in the next few minutes, both telescopes would have their shutters open within a second of each other. The clock went to six minutes past the hour. Ten seconds later Jupiter still looked quiet.

Suddenly Jupiter's limb erupted. Over the next seven seconds, both Galileo and the Hubble recorded a bright meteor appearing, brightening, and then fading away. It was S–L 9's swan song: W, the last piece, encountered its Jovian fate, brightened in a final flash of light, and vanished.

The Storm Passes

I have been one acquainted with the night.
I have walked out in rain—and back in rain.
I have outwalked the furthest city light.
I have looked down the saddest city lane.
I have passed by the watchman on his beat
And dropped my eyes, unwilling to explain.
I have stood still and stopped the sound of feet
When far away an interrupted cry
Came over houses from another street,
But not to call me back or say good-by;
And further still at an unearthly height,
One luminary clock against the sky
Proclaimed the time was neither wrong nor right.
I have been one acquainted with the night.
                                              —Robert Frost, 1928

I often think of Robert Frost's *Acquainted with the Night* when I begin an observing session. My very first was for a partial eclipse of the Sun on October 2, 1959, and by the middle of 1995 I have enjoyed more than 9400. I have indeed walked out in rain, past the furthest city light, to an observatory where I had hoped to obtain a CCD image of the approaching Halley's comet. I arrived at the observatory and waited for more than an hour while the driving rain echoed so loudly on the dome that I couldn't hear anything else. But then the sound was muted, the rain stopped, and there seemed to be a small hole in the clouds. Setting the telescope up at Halley's proper coordinates for that night, I left the dripping observatory dome shut tight. When the clearing reached the comet, I quickly opened the dome, imaged the comet for about two minutes, then shut the dome again. The wait was worth it, for I had surreptitiously captured Halley's first major outburst of a jet of dust in more than 75 years.

When I left the observatory that night, the clouds had taken over once again and I did indeed walk back in rain. On another warm night I did drop my eyes, unwilling to explain, when a police car pulled up to my parents' home at 3:00 in the morning. Having just risen in the east, Jupiter had barely cleared my neighbor's house and my telescope was pointed to it. As the police stepped down the walk, I was certain they would accuse me of spying on the neighbors. They got closer and closer, and stopped, military style, right next to my telescope. "Excuse me, sir," they said, "would you mind if a couple of nosy policemen looked at Jupiter?"

So by July 1994, could I have somehow believed that I had seen everything, experienced everything, knew it all? Not after I awoke on Saturday morning, July 23, wondering how little I really did know. What the world had seen in six days was incredible. There was a new Jupiter out there, and we would never be able to look at the planet in the same way again. So much had happened in the last week that I hadn't had the time to assimilate all the events, experiences, and changes. For Gene and Carolyn and me, the week was a siege. Privacy was a myth. "Do you have the TV on?" Mother asked Thursday evening. "Which channel?" "Any channel! The comet is on every one!"

When I woke up that Saturday, I realized that I wasn't at all acquainted with the night.

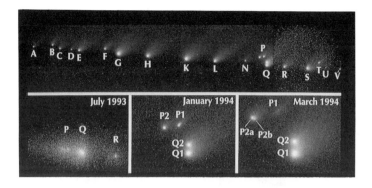

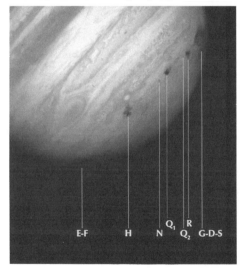

All changed, changed utterly:
A terrible beauty is born.
—W. B. Yeats, "Under Saturn"

The change from comet fragments to spots on Jupiter is possibly the most dramatic change every displayed by a member of the solar system. "From a string of pearls," Carolyn Shoemaker said, "our comet became a necklace of garnets." The top image is a version of the Hubble Space Telescope composite of Shoemaker–Levy 9, taken January 1994, in which all the fragments are lettered. The presentation is left to right. The bottom image, taken on July 22, 1994, shows several of the most visible spots on Jupiter. Top image courtesy Hal Weaver and T. E. Smith (Space Telescope Science Institute) and NASA. Bottom image courtesy the Hubble Space Telescope Comet Team and NASA.

A BOOM FROM A PLUME

Comet Shoemaker–Levy 9 was gone. On Saturday, astronomers around the world packed up their data and returned home. In Israel, Jim Scotti made time to tour the ancient city of Jerusalem before heading to England and then to his home in Tucson. Meanwhile, we drove to Goddard for a final *Daily Comet Update*. A large, exhausted group gathered at Goddard for the final press conference. As we looked at each other, we wondered if we all had just awakened from a beautiful dream, or if, after this week, we needed to go to sleep and have one. Both Hubble science teams— the one for the comet and the one for the planet—were present. Our vacuous expressions testified to our fatigue. I doubt that many of us had ever experienced such an overwhelmingly demanding stint, from the viewpoints of our scientific study and our emotions. Even the University of Maryland's SL9 message center noted the end of a stressful week: "CONGRATULATIONS AND THANKS to SL-9 observers everywhere," wrote scientists Tom Hill of Rice University and Alex Dessler of the University of Arizona, "and to the people responsible for the SL-9 Exploder, for a magnificent job. And thanks to SL-9 (R.I.P.) for a magnificent show."[1]

It was time to pick up the pieces. What had happened? What did we learn? In the six months after the impacts, few of the hard questions would really get answered. Scientists would be mulling over the mother lode of data for years. In the great forest of Jupiter, a reporter asked, did the comet impacts make a sound? According to Caltech's Andrew Ingersoll, they probably did.

How would astronomers know whether or not the impacts actually generated an explosive sound? The evidence came initially from the circular formation around the G impact site. Initially, scientists thought it was indeed a sound wave. This preliminary interpretation would be changed as time went on, and it turned out that the wave was propagating much too slowly to be a sonic wave. Just how the wave was being generated, and from how deep in Jupiter's atmosphere, was a question that would not be answered so quickly.

THE WATERBED WAVE

"Boom from a plume" was catchy enough as a title for a press conference, but by the end of the year Andrew Ingersoll had rejected the sonic wave theory to explain the waves. The Hubble, it turned out, recorded rings around five of the impact sites—A, E, G, Q1, and R. First recorded from 1 to 2½ hours after the impacts, each ring expanded. Soon after impact week, Ingersoll realized that the rings were not moving out fast enough to be sound waves. Since the speed of sound in that part of Jupiter's atmosphere was 775 meters per second, these waves were propagating much too slowly for that.

If sound did not generate these rings, what did? The ring speed was constant in all five samples. Not only was the velocity of the ring known, but their temperature was measured as being between 253 and 335 Kelvin. Armed with these data, all Ingersoll had to do was to find a model that fit. He assumed that on Jupiter the abundance of water—or the ratio of oxygen to hydrogen— was about 10 times that of the Sun. If this assumption were correct, then the observed wave seemed to be propagating horizontally while trapped in a stable moist layer. It would be similar to the wave you get if you jump on a waterbed.

Other questions arose: A reporter from Reuters asked if any of us were feeling postimpact letdown. Gene allowed that although the comet was gone, it was hard to be sad since it had performed so well. Ingersoll added that he used to have a "post-Voyager letdown" after each of the encounters that Voyager made with Jupiter, Saturn, Uranus, and Neptune. "I'm sure I will miss these good old days!" he admitted.[2] As would we all.

During Friday's *Daily Comet Update*, an alert reporter noted that the IUE had been observing Jupiter 24 hours a day, but that the Hubble was also looking at other targets during this time. He asked why this was the case. "The people closest to Hubble have *no* complaints!" Andrew Ingersoll emphasized. Besides the fact that for half of each orbit, the Hubble could not see Jupiter because Earth was in the way, the telescope's observations of Jupiter and the comet formed the basis of a magnificent data set that, Lucy McFadden insisted, was richest in its combined form.

The Hubble's performance led off the final day's press conference, which began with the entire Hubble comet–Jupiter science team being asked to come forward to be congratulated. The kudos were richly deserved. Many observatories had contributed spectacular photographs, but the wildest praise in the press was generally reserved for the Hubble Space Telescope and the many people who worked on it. Part of the reason was the "comeback kid" image that the Hubble has generated since its repair. "The fact that the Hubble Space Telescope was barely noticed last week is a testament to how well it's working," went an editorial in *Space News*. "For six spellbinding days millions of people worldwide were captivated by the startlingly clear images of comet Shoemaker–Levy exploding into Jupiter's atmosphere. The most spectacular of all were the Hubble pictures.... It was a great week for science and an appropriate way to celebrate the 25th anniversary of the moonwalk."[3]

The behavior of the A impact site—the "little pediddle" as Reta Beebe had compared it to the later, larger impact areas—became a subject of considerable excitement this last day. The latest Hubble image showed the dark material apparently developing a small "curlicue tail" stretching toward a white oval storm nearby. Winds from the storm were affecting the evolution of the A site and scientists had a new tracer for circulation in Jupiter's upper atmosphere. We also got a second look at the area where spectra had revealed the emergence of sulfur a few days earlier. The spectrum had changed radically, as more material from the comet—magnesium, silicon, and iron—had made an unexpected appearance.[4]

Meanwhile, an observation came in from Jupiter—actually from the town of Jupiter, Florida. Michael Palermiti sent NASA an image that showed the G and L impact sites looking very prominent through a 20-inch reflector with CCD attached. Steve O'Meara from *Sky and Telescope* magazine reported seeing the G site using a 60-millimeter (that's a 2.4-inch diameter) finder telescope.

Many amateur astronomers photographed Jupiter during impact week. This one is a part of a series taken by Michael Palermiti, who observes from Jupiter, Florida. Palermiti used a 20-inch Cassegrain reflector telescope and CCD to obtain this image.

But time was passing. Gene looked at his watch, and announced that he and Carolyn had to leave to catch their flight home. Both Don Savage and Gene asked Carolyn to name her favorite memory of the week: "David and Gene and I all three feel like proud parents," Carolyn answered, "because this comet, this child, has lived up to all the great expectations and the great desires we had for it. And now it's not just our comet; it belongs to the whole world. And the science team is going to produce some absolutely great things in the days to come, and so are the amateurs, and so are all of those who observed it. We are deeply grateful to all of you."

23 July 1994   00:23 UT   23 July 1994   00:31 UT   23 July 1994   02:16 UT

Don Parker took these three images of Jupiter using his 16-inch telescope and CCD on July 23, 1994.

COMET VS. ASTEROID

Exeuent Gene and Carolyn, but the questions went on, some reflecting a growing interest in S–L 9's nature. Had it been a comet or an asteroid? When the question was first raised at an earlier press conference, I held that the distinction should be made by observation, as it has been for almost two centuries. The word "comet" comes from a Greek word for long-haired star, perhaps a bright furball. The more recent term "asteroid" derives from a Greek word that suggests a starlike appearance. If the newly discovered object was fuzzy, it was a comet, but if it was starlike, then it was an asteroid.

There are causes for this difference in appearance. Generally comets release dust and gas that forms a coma and tail, causing the fuzzy appearance. Asteroids do not. But if a comet is so far from the Sun that it is completely quiescent, it can resemble an asteroid.

For some reason, the idea emerged during these press conferences that comets must emit gases, or else they should be called

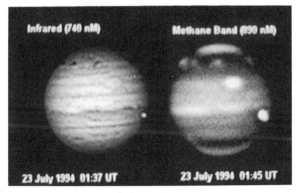

Compare these infrared images of Jupiter, taken by Don Parker on July 23, 1994, with the visible light images in the preceding figure.

asteroids. Actually, asteroids and comets are a continuum of objects. Comets do not produce comae during their full journeys round the Sun; when they don't, they look like asteroids and are so designated if someone discovers them in that state.

As far as we know, the orbital history of Shoemaker–Levy 9 suggests that it was cometary as well. We know of several examples of comets that have encountered Jupiter, but so far we know of no asteroids that have done this. According to Brian Marsden, before Jupiter captured the comet earlier this century, Shoemaker–Levy 9 was probably in a cometary orbit about the Sun with a period of about eight years and at a distance some four times greater than that between Earth and Sun (we call that four astronomical units.)[5]

CLEANING UP

After a good night's sleep, observers took a fresh look at their data and found some interesting new things. "In re-examining our Jupiter data," wrote the observers from the 200-inch at Palomar,

The changing view of Jupiter is easily seen in this set of Don Parker images taken on July 24, 1994 using his 16-inch telescope and CCD.

"we have identified what appears to be a prompt flash from the V impact on July 22.... The image is point like.... Impact sites H, Q1, R, G, and L are visible in the same frame.... There is little doubt that this outburst is related to the collision of S–L 9 with Jupiter because our pre-encounter runs in April and May show no deviations larger than 5 percent at 21, 18, and 9 cm during one month observation."[6]

Other types of observations were revealed as time went on. A pronounced increase in Jupiter's microwave radio emission, for example, was reported by M. J. Klein and S. Gulkis using Jet Propulsion Laboratory's Goldstone tracking antenna. They first noticed the increase on July 17, and it peaked on July 23.[7]

For me, the impacts ended on a sad note. I talked over the telephone with Beulah Kushner, my mother's sister and one of my favorite relatives, who had been very ill at her home in New Orleans. She had watched the television all week, reveling in her nephew's frequent appearances. Even though I could hardly hear her, I found her more enthusiastic than she had been for a long time. I was very fond of Aunt Beulah, and talking with her on that

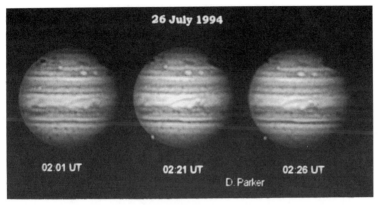

Jupiter on July 26, 1994, imaged by Don Parker using his 16-inch telescope and CCD.

July day seemed a fitting way to end this momentous week. But since she passed away soon after, it was the last time I spoke with her.

I left Washington half exuberant and three-quarters exhausted. My mother and I went to the airport and caught separate flights leaving at almost the same time—hers for Montreal and mine for Tucson. Once home, I had a few days to catch my breath before heading up to Kitt Peak for a new phase of the Shoemaker–Levy 9 program.

The spectacular duet of Shoemaker–Levy 9 and Jupiter was over, but it was not over. The comet's trailing wing continued to encounter Jupiter for more than a month, and its pieces, which could be as large as houses, would be striking on the side of Jupiter facing Earth. Although the main observing campaigns were concluding, several groups were continuing to watch the planet in hopes of catching evidence of further impacts. A group of us from Tucson's Planetary Science Institute (PSI) searched for small impacts from larger members of the trailing wing. During our two nights on Kitt Peak, the weather did not cooperate. Jupiter

Jupiter on July 27, 1994, imaged by Don Parker using his 16-inch telescope and CCD.

was only intermittently visible through the clouds on the first night, and we saw no obvious flashes or increases in brightnesses on the planet's limb.

The collision of S–L 9 seemed a particularly fortuitous event for PSI, whose scientists have considerable experience both in constructing theoretical models of how comets are constructed, and in studying the dynamics of collisions. Eileen Ryan, for example, has studied the distribution of fragment sizes from many possible conditions of impact. The main goal of PSI's study of S–L 9, Ryan notes, "was to analyze the size distribution of the fragments by observing (at KPNO) impact flashes from the available known fragments, as well as from undiscovered smaller fragments that may be plentiful." Ryan added that their initial reduction did not detect any additional small fragments.[8]

Other groups had more luck. Out in California, one week after the last impact, Phil Kesten and a group of Santa Clara University students were observing on July 29 with a CCD though a 16-inch reflector and saw an area brighter than Io surrounded by a dark region, close to where the material in the comet's west-

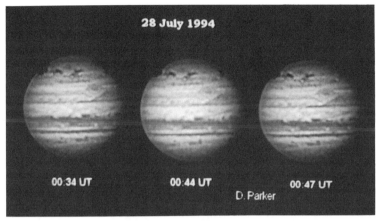

Jupiter on July 28, 1994, imaged by Don Parker using his 16-inch telescope and CCD from his home in Coral Gables, Florida. While climbing down the ladder after taking these images, Parker injured his knee so severely that he needed surgery and was on crutches for months.

southwest trail should be hitting. They interpreted this as a real event, possibly the plume from a small comet fragment.[9]

PAINTING A PICTURE

By the end of 1994, scientists were stymied on several key questions. How big were the comet fragments? What were they made of? How deep did they penetrate on impact? One day before the first hit, most scientists thought that if we could observe the impacts, we would have the answers easily. This was not the case.

In the short term at least, astronomers seem cursed by the sheer volume of data. To try to make sense of the many datasets, in November, Torrence Johnson, project scientist for the Galileo spacecraft, called a meeting at the Jet Propulsion Laboratory. The original idea was to evaluate observations made only of fragment

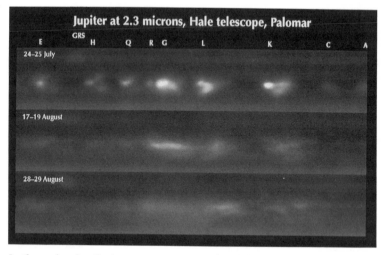

In the weeks after the impacts, astronomers using Palomar Observatory's 200-inch (5-meter) telescope followed Jupiter carefully. The top row was taken a few days after the end of the collisions; the middle row shows the impact region 3½ weeks later. The bottom row displays the region five weeks after the end of the impacts. Courtesy Philip Nicholson and Gerry Neugebauer.

G's impact by the Galileo spacecraft, the Hubble Space Telescope, and several ground-based telescopes. When I arrived at this "G Meeting," I was surprised at the number of people—more than 30—who crowded the room. At this meeting some fascinating news emerged, and not just on fragment G. It turned out that both the Hubble Space Telescope and Galileo were imaging Jupiter at the same time during the G impact, and that a Hubble image, using a green filter, was taken within a single second of Galileo's shot of the W impact at its brightest. Both Galileo and HST, it turned out, were observing W coming in as a bright meteor over Jupiter.

In October, several hundred planetary scientists—probably more than half of all planetary scientists in the United States—met in Bethesda, Maryland. At this meeting of the American Astronomical Society's Division for Planetary Sciences, astronomer Andrew Ingersoll polled the scientists on how they felt about major questions relating to the impacts. Most of the scientists who responded to Ingersoll's questionnaire had been involved in the event in some way, either by observing the impacts or by modeling them both before and after the events.

This exercise demonstrated an interesting point: science is not inherently a democracy. While the vote was taken with a certain spirit of humor, the responses reflect only a sense of where we were four months after the impacts.

Was S–L 9 a comet or an asteroid? A substantial majority of the scientists answered that they would describe S–L 9 as a comet, not as an asteroid. The fact that the Hubble images of our string of pearls always showed a thickening amount of coma near the nuclei convinced most astronomers that dust was being produced all the time—a dead ringer for a comet. Asteroids do not generally produce dust.

Where did Shoemaker–Levy 9 originate? A majority of scientists answered that they suspect that the comet began its wanderings in the outer solar system, and that its capture by Jupiter came at the end of a long journey. Only a few thought that the object originated near Jupiter or in the main asteroid belt between Jupiter and Mars.

What were the dark spots made of? The impacts produced large, dark spots that were easily visible through small telescopes. A slim majority thought that they were comprised of hydrocarbons, of which soot is a familiar form. The others thought they were made up of silicate material.

The spectra of several metals were revealed during the impacts. Where did the metals come from? Almost everyone thought that the metals derived from the comet, and not from Jupiter. On the other hand, more than half of the scientists thought that the sulfur that the Hubble telescope detected was from Jupiter.

How large were the largest fragments? Four months after the explosions, astronomers were still uncertain how large the frag-

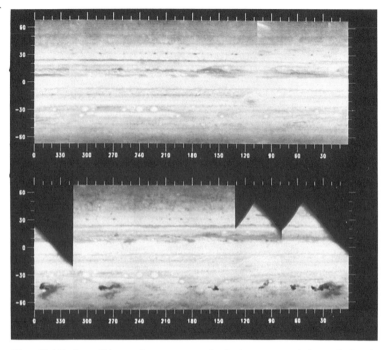

These Hubble Space Telescope images contrast Jupiter's appearance before and after the impacts. The preimpact view is a composite of several imaging sessions on July 15, the day before the first hit. The second view was taken in similar fashion on July 23. Courtesy Hubble Space Telescope Comet Team and NASA.

ments were, or how massive the resulting detonations were. Possibly the range for the latter was from 10^{27} to 10^{29} ergs (20,000 to 2,000,000 megatons of TNT). In one model, a comet only 600 meters (2000 feet) across, hurtling into Jupiter at 60 kilometers per second (134,000 miles per hour) could produce the largest plume and all the effects associated with it. Other theorists disagreed. Three kilometers (2 miles) across, one group suggests, is a more realistic size for the largest comet fragment. A third group opted for 2 kilometers (a little more than a mile) wide.

For the sake of life on Earth, we should hope that the comet fragments were large. If the comets were at the small end of the size range, then the Earth has much more to fear than we previously thought from collisions with small comets.

How were the fragments put together? Scientists were almost equally split among three possibilities. The comets might have been rubble piles of loose, unconsolidated gravel. Fluffy snow, with a mixture of frozen gases and dust, was a second possibility. Finally, they might have been coherent, solid asteroidlike masses.

How deeply did the comets penetrate into Jupiter's atmosphere? Suggested answers were, to say the least, diverse on this question, although new evidence emerged at the meeting that the largest ones might have survived while plunging more down more than 300 kilometers (about 200 miles) below the cloud tops.

What is the frequency of events of this magnitude on Jupiter? Although most impact experts thought that we were seeing a once in a thousand year event, the fact that this comet performed a two-act play by splitting up in an earlier encounter with Jupiter and then colliding with it, makes this type of event even more rare. We may never know how privileged we were to see so many comets strike Jupiter in just a few days, but the fact that dark spots of this number and intensity have never been seen on Jupiter since the invention of the telescope is good evidence that this was a rare event.

Surprisingly, a few scientists voted that something similar to what we saw might actually occur every century on Jupiter. If this is true, why has no one noticed such an impressive series of dark spots before?

There were two interesting developments by the end of 1994. A careful look at the data obtained by the Hubble Space Telescope from several plumes showed that they all rose to the same height. This startling conclusion derived from fragments as faint as A to those as bright as G. No one was trying to say that this evidence meant that the fragments were identical in size. The G impact site was much larger and long-lasting than A was. Then, Galileo detected the violent end of fragment N—the flash of one of the faintest fragments—and found it fully half as bright as K!

Gene, Carolyn, and David (right to left) in the small room beneath the 18-inch telescope on their last night at Palomar Observatory, December 1994. Photograph by Jean Mueller.

JUPITER RETURNS

By the end of October, Jupiter was setting so soon after the Sun that observations became impossible. Would any evidence of the impacts be visible after Jupiter reappeared in the morning sky?

In early December, Gene and Carolyn and I met with Henry Holt for our last observing run at Palomar. Although we were sad that our Palomar observing program was coming to an end, we felt that it was time for something new. Our comet and asteroid search program had run its course, and the plan was to replace it with two new projects. The Lowell Observatory Near-Earth Object Search, or *LONEOS*, is one program. Using a 24-inch Schmidt camera with CCD attachment, we would expand the search for asteroids and comets considerably. Our other program uses two

It is hard to describe the utter jubilation felt throughout NASA at the end of impact week. Peter Jedicke photographed this sign at Goddard Space Flight Center.

8-inch-diameter Schmidt cameras, mounted together but covering adjacent areas of sky.

As we left Palomar for the last time, we knew that our favorite planet was lurking out there somewhere in the sky near the Sun.

DECEMBER 17, 1994

On December 17, 1965, I first began looking for comets through the bright sky over Montreal. At the end of 1994 I celebrated 29 years of searching, a labor of love that led to eight visual discoveries and 13 photographic ones shared with Gene and Carolyn Shoemaker. The night of December 17, 1994, was clear and transparent, but a bright Moon ruled out any searching for comets. Nevertheless, Jupiter was now far enough from the Sun in the predawn sky that I hoped to see the planet again. I waited for my rendezvous with Jupiter, and just before dawn I could see the

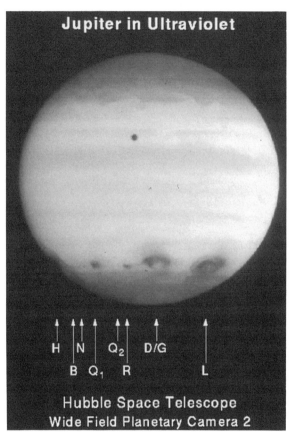

By the end of impact week, so many spots had appeared on Jupiter that it was getting hard to keep them straight. This Hubble Space Telescope image shows a large group of sites in violet light. Courtesy Hubble Space Telescope Comet Team and NASA.

planet peek above a distant mountain in the southeast. I set up my 8-inch reflector and peered through the eyepiece. The seeing was so poor that I could not even make out that Jupiter was round.

By 7 a.m. the sky was quite bright, and I looked through the eyepiece again. Although the seeing conditions were still poor, I could see Jupiter looking more like its usual self. But the sky was brightening so quickly that I was worried that the planet would simply fade away before the atmospheric steadiness was good enough for me to try to detect the spots. Moreover, five months had passed since the impacts. There quite probably was nothing left to see.

I peered through the eyepiece again and tweaked the focus. It was as though I was looking at Jupiter from the bottom of a pond, but at last the water was getting steadier. Then for one brief moment—the shortest moment—Jupiter was there. Rock steady. I saw the dark band of the North Equatorial Belt, and around the southern hemisphere—at the very latitude where the impacts took place—was a dark belt crossing the whole face of the planet.

I couldn't believe my eyes. And at first I didn't believe them. What a present this would be for the 29th anniversary of the start of my comet search! I waited for the next steady moment and spied the belt again.

The seeing was now good enough that I could make out detail on Jupiter's new belt. There was actually a thickening of dark material not far from the planet's west side, which I took to be one of the major impact sites, perhaps G or L. I watched as long as I could, but with the Sun about to rise, I had trouble making out any detail on that distant world whose return to the night sky I had just witnessed.

I went indoors and typed this message:

This morning from 13:40 to 14:00 UT I observed Jupiter visually with my 20-cm f/7 reflector, through poor seeing. The dark material left from the impacts of S–L 9 is still there, and still obvious. Crossing almost across the correct latitude is a dark bar, which widens into a large spot about a third of the way from the western limb. With the exception of the North Equatorial Belt, these features are the most conspicuous on Jupiter.

High in Jupiter's atmosphere, the dark material had persisted for five months. It was so thick that the regions below were receiving a vastly reduced amount of sunlight. Imagine a similar bombardment taking place here on Earth. Its atmosphere covered by dark material for so many months, our planet's biosphere would be threatened by a frigid temperatures, a shutdown of photosynthesis, and starvation. The great legacy from Jupiter and Shoemaker–Levy 9 could be that once scientists are able to interpret the events, we will have a better idea whether Earth could handle such an awesome fate.

The Day I Heard the Moon Roar

Twice each day, at the eastern end of Canada's Bay of Fundy, a flow of water equal to the combined currents of all the rivers on Earth passes through the Minas Channel into the Minas Basin. I had a chance to watch that happen during the first week in May, 1995. I was on a journey to two places—Acadia University in Wolfville, Nova Scotia, to accept a doctor of science degree, and to Baltimore, Maryland, to attend the meeting on Shoemaker–Levy 9 that would provide material for this final chapter. On Thursday, May 4, Roy and Gertrude Bishop and I hiked out from the foggy fishing village of Scots Bay along the top of North Mountain. An Acadia physics professor and experienced astronomer, Roy knew the significance of the tidal flow we were about to see, and I was glad to be with him. Our destination, Cape Split, was two and a half hours away. As we approached it, the sound of the water grew louder. Suddenly the forest stopped. We were on a small grassy area at the edge of a two-hundred fifty-foot precipice of basalt. I looked down and saw the equivalent of all the rivers on Earth flowing past. The Minas Channel is a tough pipe for all this water: The cliffs on both sides, as well as the underwater base, are built of solid basalt that formed at the start of the Jurassic period.

We arrived at our true-life Jurassic Park just as the incoming tidal flow was at its maximum, and as we munched on sandwiches, 10^{12} gallons—a million million gallons—of water poured into the Minas Basin. Meanwhile, the first quarter Moon hung above us in the sky. It was near apogee, its farthest point in its orbit around the Earth. Still, that small body, from a distance of almost 400,000 kilometers, was responsible for the tremendous noise in the channel below as the water rushed by. After a few hours the current slowed, then stopped. For a half hour or so the water around Cape Split was still. Then the huge basin started to empty, flowing *in reverse*, just as if the place were a movie running backward.

On the return hike we talked about how what we had just seen related to Comet Shoemaker–Levy 9 when it was passing by Jupiter in July of 1992. We knew that the water and noise below us was a special case involving resonance. The Moon's role, helped about a quarter as much by the Sun, was driving tides throughout the Atlantic Ocean. The actual amplitude of the open ocean tide is not that great, perhaps about 1 meter. Textbooks describe the tides as due to the differential gravitational pull of the Moon (and to a much lesser extent, that of the Sun) across the Earth's diameter. Out by Jupiter on July 7, 1992, the incredibly fragile Comet Shoemaker–Levy 9 was now undergoing the same type of resonance as it tried to edge its way past Jupiter.

However, the Atlantic tides drive the tidal oscillation in the Fundy system, which includes the Gulf of Maine, the Bay of Fundy, and the Minas Basin and its network of rivers. If you push a child on a swing, giving a little push in step with the motion of the swing, you can generate a much higher and more adventurous trip; the force of your hand is in perfect resonance with the motion of the swing. Similarly, the Moon is only indirectly responsible for these high tides; it drives the Atlantic tides (your force on the swing), which are nearly in step with the natural oscillation of the Fundy system. This "resonance" phenomenon greatly amplifies the tides, particularly in the Minas Basin.

May 4, 1995. This view from Cape Split shows the awesome rush of water at mid-incoming tide. In three hours the water will be still, and in six hours the water will be flowing swiftly in the opposite direction. The peaceful grass is deceptive, I stood atop a narrow column of rock at the edge of the precipice to get this shot.

How can we stand on Cape Split and compare the tides on Earth to the force that tore Shoemaker–Levy 9 apart? By taking three steps. In step 1, we would move the Moon to the same distance that S–L 9 was from the center of Jupiter, which was 21,000 kilometers plus Jupiter's radius of 71,000 kilometers, for a total distance of 92,000 kilometers. As a result, the tides on Earth

Jupiter's enormous gravity is fully capable of tearing a comet apart. Even the weak gravity of the Moon, at a distance of 240,000 miles, is capable to causing dramatic tides on Earth. The highest tides occur in Minas Basin, at the head of the Bay of Fundy in Nova Scotia. Roy Bishop, a physicist at Acadia University, took these two photographs on January 29, 1979. They show low tide at 8:30 am and high tide only five hours later, at 1:30 p.m.

The 16-meter (more than 50-foot) tide was higher than normal for four reasons: The Moon was at perigee, at its closest point in its orbit of the Earth, only the day before. The Moon was also at its new phase, so it was in line with the Sun, which also exerts a tidal pull. The barometric pressure was low. Finally, the Earth itself was as near its closest point to the Sun. Bishop knew the tide would be dramatic, but it rose so high that he had to step dangerously across a fully submerged beam to get the second picture.

would be 70 times stronger. The tidal roar at Cape Split would be completely deafening as many times the amount of water stormed through.

Step 2 would require us to replace the Moon, at its new close distance, with Jupiter. This step would increase the tidal pull by *26,000 times* from what we saw on May 4. When we multiply this by the 70-fold increase in step 1, the new tidal stress would now be *1.9 million* times the force of May 4. Being well within Jupiter's Roche limit, the Earth would start to tear apart. Soon our home planet would become a bright ring around Jupiter.

But just before that happens, we could invoke step 3, where we'd replace the Earth with Comet Shoemaker–Levy 9, an object at least 1.5 kilometers in diameter. Because the comet is so much smaller, Jupiter's tidal stress would be 8,500 times smaller than it would have been on the Earth. We have to divide 1.9 million by 8,500, to get 220.

Our conclusion: Comet Shoemaker–Levy 9 was stressed about 200 times greater than the Earth was on the day I stood on Cape Split. For the first time, I could empathize with the comet's struggle to get past Jupiter on July 7, 1992—a struggle that it could not possibly win.

I thought about the implications of this as I flew on to the Baltimore meeting, where scientists were presenting their latest findings from both observational and theoretical study. The information that follows in this chapter is based on what happened in Baltimore, at Colloquium 156 of the International Astronomical Union: *The Collision of Comet Shoemaker–Levy 9 and Jupiter*. Sponsored by the Space Telescope Science Institute, the meeting was held on the parklike grounds of the Johns Hopkins University.

WHERE IT CAME FROM

Based on accurate measurements of the positions of the fragments, especially fragment K near the center of the comet train, and of the impact times of that fragment, Shoemaker–Levy 9's orbit is about as well known as it is going to get. If in the future someone finds images of a faint parent body on a photographic plate taken before July 1992, we could then have a much better idea of the comet's path. Paul Chodas and Don Yeomans have calculated the orbit backward in time and have concluded that the comet probably began its dance about Jupiter in 1929, the year of a different kind of major crash that affected almost everyone on Earth. It was also the year that Carolyn Shoemaker, who first recognized the image of Comet Shoemaker–Levy 9 on our photographic film, was born.

In all likelihood, Shoemaker–Levy 9 began as a Jupiter family comet like several that are discovered each year. Most of these comets orbit the Sun in periods of from half a dozen to a dozen years. (Gene Shoemaker estimates that there is a three-tenths of 1 percent chance that the object was a carbonaceous, or "C-type," asteroid that had somehow fallen into this orbit.) A few of these comets, like Helin–Roman–Crockett and Gehrels 3, temporarily go into brief orbits about Jupiter for a few revolutions before returning to orbit about the Sun. We did not know about either of these comets when they were in orbit about Jupiter; in fact S–L 9 was the first one caught in this act. "Perhaps the best evidence that SL9 was a comet," wrote Hal Weaver, "was the existence of a persistent circularly symmetric coma around each fragment, possibly indicating continuous dust production."[1] If all the dust had been produced at the original breakup, the coma would not have remained so spherical until a few hours before impact.

WHAT WAS THE COMET'S DIAMETER?

How large was the original comet, and how big were the impactors? These were the first questions occupying the thoughts of scientists. Since we knew precisely how fast the fragments were coming in, the amount of energy Jupiter received from the impacts was directly proportional to the cube of their size. Most of the computer models generated by theoretical astronomers were starting to agree in several respects: that the original comet was about 1.5 kilometers in diameter, that the bodies that created such dramatic effects on Jupiter were never more than a kilometer in diameter, and that the total energy released was 10^{28} ergs, or about a quarter million megatons. Even our Hubble Space Telescope team, whose data suggested that the fragments could be as large as 4 kilometers on the largest impactors, considered that the actual diameters might be much smaller.

The Hubble Space Telescope's observations of Shoemaker–Levy 9 began on July 1, 1993, and lasted until July 20, 1994.[2] The telescope's view of the impacts revealed major atmospheric changes, including plumes rising to 3300 kilometers at velocities of about 10 kilometers per second.[3] When the telescope looked at the comets before they hit, it could not resolve anything smaller than 250 kilometers (close to 150 miles) at the distance of Jupiter and the comet. However, the observers unsuccessfully tried to detect a telltale point source of brighter light that would definitely be from the nucleus. "We conclude only," the Hubble Observing team writes, "that the larger fragments may have been a few kilometers in diameter, but that smaller values cannot be ruled out."[4] A year before the impacts, the first attempt to provide an answer was by Jay Mellosh and Jim Scotti, who proposed that the proginator could not have been more than 2.2 kilometers across, and now, considering that the comet got closer to Jupiter than they had calculated at the time, they have revised their estimate comet's diameter down to 1.5 kilometers, or about a mile.

Mordecai-Mark Mac-Low, from the University of Chicago, had a model that explained virtually all the impact features, especially the length of the chain of fragments, and the number of fragments, the dimness of the fireballs, and the undetected seismic waves, with a progenitor as small as 1.5 kilometers. Kevin Zahnle of NASA's Ames Research Center agreed that the length of the comet chain precluded a comet progenitor larger than 1.5 kilometers; "Keep them small and keep them high" was the theme that seemed to fit the observations. Thus fragment G, Zahnle maintained, was between 600 and 700 meters (about a half mile) across, and R was about 500 meters across. Michael Mumma, from Goddard Space Flight Center, came up with a similar solution, as did Terrence Rettig of the University of Notre Dame. Most modelers agreed that the comet fragments were not thick balls of ice, but a swarm of different sizes of material of different compositions, like rice pudding or tapioca with raisins.

From Sandia National Labs outside Albuquerque, Mark Boslough and David Crawford suggested that large fragments like G could have been as large as 2 kilometers across. However, they attached a caveat: This larger figure could be the total diameter of a cluster of smaller fragments, or clouds of dust and debris.

Still arguing for a big progenitor were Zdenek Sekanina, Donald Yeomans, and Paul Chodas of NASA's Jet Propulsion Lab. They used a different approach to come up with a progenitor as large as 10 kilometers. The time of the "dynamic" breakup occurred some 2 hours *after* closest approach to Jupiter, Sekanina maintained in Baltimore, and that time controlled the length and position angle, or orientation, of the train. Also arguing in favor of a larger size was the comet's behavior; the parent did not originally break up into 21 pieces, but into 8, and other pieces broke off at times after the initial July 7 breakup. This secondary fragmentation, Sekanina insists, has two implications: that possible cracks occurred at the original tidal split, and that the dust clouds surrounding the fragments remained symmetric until hours before their impacts.

As interesting as the impactor size was the graded distribution of material that made up the fragment. "There was a zoo of different kinds of nuclei out there," said Gene Shoemaker. "The comet was probably a pile of debris formed by ancient collisions among its parts." Counting the vanished fragments J and M, altogether there were 23 nuclei whose collisions could have been seen. Observers apparently have not confirmed seeing the impacts from fragments J, M, and V.

Whatever the correct figure may be, the comet's size has profound implications for Earth. The conventional wisdom is that if a comet or asteroid at least a kilometer in diameter were to hit the Earth, the consequences would affect the entire biosphere. But the largest fragments of S–L 9 left dust clouds that grew larger than the whole Earth an hour after impact. The entire impact latitude of Jupiter was covered with a dark cloud that lasted for more than a year. If these bodies had hit Earth, the scenario would have been similar to a nuclear winter (but without the radiation). The dust cloud they formed would have covered the planet, thickly, and the Earth would have cooled off drastically over a period of months or years. The darkness would have shut off photosynthesis, thus eliminating a year's growing season and causing worldwide starvation.

Since an asteroid hitting Earth would be moving some three times slower than the comets were moving on Jupiter, the 700-meter-diameter object hitting Jupiter would be the equivalent of an asteroid of about 1.5 kilometers diameter hitting us. However, a comet coming in from a great distance could well hit us at the same speed as S–L 9 struck Jupiter. Were Comet Swift–Tuttle, the parent of the Perseids, or Tempel–Tuttle, the progenitor of the Leonid shower, to strike, they would be moving at close to the 60 kilometers per second of S–L 9's speed.

The fossil record on Earth records the great extinctions of the past, but it does not note suffering. If the impact of a body only half a kilometer across caused mass starvation on the planet 10,000 years ago, we might not even know about it today. Our human ancestors would have suffered greatly, but the sudden, and short-lived, climate change, would have left no record. If the crater were on an ocean floor we would not even know of its existence.

WHAT HAPPENED AS THE FRAGMENTS HIT

Any attempt to model what happened on Jupiter would have to account for several observed phenomena. These include how all the plumes rose to the same height of 3,300 kilometers above Jupiter's cloud tops, how they rose and collapsed, and how their material formed the dark inner clouds and then flowed laterally across Jupiter to make the crescent-shaped outer clouds.

An approach proposed by Gene Shoemaker, Paul Hassig of the Titan Corporation, and David Roddy of the U.S. Geological Survey, divided the problem into two phases. The first was the impacts themselves; the second was the planet's reaction to the impacts. Each major impact cloud, they note, consisted of three zones: a dark inner core with material dredged from the ammonium hydrosulfide layer, and an outer dark crescent, with a transparent region in between. As each plume collapsed, its base developed a "skirt" that expanded laterally and became a crescent-shaped cloud. The crescent shape resulted from the way the plume expanded out in the "backfire" direction—out the same entry "tunnel" that the incoming fragment had dug into Jupiter's atmosphere. The crescent did not lose its shape for quite some time, an indication that it formed high in Jupiter's stratosphere, where the winds are relatively calm. The core cloud, on the other hand, deformed quickly. Its base was probably fairly low in the troposphere and might have been a deep column that began as deep as the ammonium hydrosulfide layer of Jupiter's clouds (a layer that lies hidden just below the top ammonia cloud deck) and then towered some distance into the stratosphere. The column consisted of a combination of cometary material and material from the ammonium hydrosulfide layer. The transparent region in between, they believe, consists of different material. The middle transparent layer is made up of material from the ammonia cloud layer. The outer crescent is a dark brown aerosol of cometary material that has reacted with heated gas in the planet's atmosphere. Also, the G crescent had 16 easily identified rays in it, which, they believe, reflect the fragment's breaking up into 16 large pieces as it hit.

In Kevin Zahnle's model, all the energy forced its way out the top, in a mighty back flush that spewed out material on the ballistic trajectories high above Jupiter's atmosphere. Zahnle modeled the luminosity of the impacts at different wavelengths in order to account for the total observed infrared radiation. But Mark Boslough and David Crawford suggested the comets penetrated deeper. Each fragment began to break up as it hit the 1 bar level, or the top of Jupiter's clouds, and continued breaking up as it hurtled down another 300 kilometers, sloughing off material as it went down. We never saw the manifestation of the material that went further down, Crawford explains. We only saw the highest material.

UNDERSTANDING THE COMPLEX LIGHT CURVE

The impact events were so complex, and were observed by so many instruments in space and on the ground, that figuring out which telescope saw what was one of the biggest challenges we faced during the year after the impacts. The breakthrough finally came during the afternoon of the Baltimore meeting's second day. Clark Chapman had been trying to bring together all the observations for months, and with the help of others he finally figured out a chronology for three of the impacts—H, K, and L. To this description I have added information gleaned from other sources:

First Precursor Episode

Stage 1. Hailing down on Jupiter was a central fragment, or possibly a swarm of fragments, hundreds of meters across to several kilometers across. It was surrounded by a coma of dust and rock to bolder-sized debris several thousand kilometers long: by the time of impact, the coma would have stretched out to great lengths. The entire complex was heading for impact at a 45-degree angle to the local vertical at a velocity of 60 kilometers per second. The first contact was with the huge coma of dust. As these small particles encountered Jupiter's upper atmosphere, there was a slow, then vigorous, rise of light that lasted as long as a minute. This first precursor was not seen by Galileo or by the Hubble.

Stage 2. As the coma continued to react with Jupiter's upper atmosphere in a tremendous meteor shower, the central fragment vanished behind the edge of the planet. As seen from Earth, the light level dropped suddenly. Hubble and Galileo have images taken almost simultaneously of this phase of W.

Stage 3. From Earth, as we lost sight of the incoming fragment, the brightness fell dramatically. And as most of the coma disappeared behind the edge of the planet, its brightness continued to drop, but less dramatically, for about a minute.

The Meteor Episode

Stage 4. In each of the H, K, and L impacts, as the brightness dropped, there was a small momentary increase in brightness about five seconds after the peak of the first precursor.

Stage 5. From the ground this was barely detectable, but at the same instant, two of Galileo's instruments recorded a sudden jump in signal strength.

Stage 6. Five seconds after it began, the brightening reached its peak, as seen by Galileo's solid-state imaging system. This, Chapman concludes, was the explosion of the meteor as it struck Jupiter's lower stratosphere. It should not have been visible at all from Earth-based telescopes. The fact that it was (stage 4) indicates that the bolide reflected off high-altitude dust. With fragment K, the brightness fell off sharply before increasing again over another 10 seconds.

The Fireball Episode

Stage 7. A big explosion occurred at the top of the meteor trail as the fireball started to rise and expand. For 15 to 45 seconds, Galileo's nonimaging instruments recorded this event.

Stage 8. A full minute has passed since the first precursor reached its peak. Ground-based observers now recorded a second precursor, an abrupt surge in brightness, 10 times brighter than the first one, as the mighty fireball, with temperatures between 7000 and 12,000 degrees Kelvin, suddenly revealed itself from behind the edge of the planet.

Stage 9. The peak of the second precursor. There was about a one-minute time lag between the first and second precursors, which Chapman explains by the fireball's speed of ascent, at a 45-degree angle, of 17 kilometers per second. The temperature at the top of the fireball was an astonishing 18,000 degrees Kelvin.

Stage 10. The decline in brightness was slow and didn't return to its original level. Again, this second precursor was observed only from the ground.

The Main Event: Ballistic, or Infalling Debris Episode

Stage 11. As millions of tons of material rained down on Jupiter's stratosphere at speeds faster than 5 kilometers per second, Hubble recorded the flattening plume. This phase might have been barely recorded by Galileo's NIMS (Near-Infrared Mapping Spectrometer). Observatories on Earth with infrared cameras, however, recorded the surge of brightness as an intense heating of the upper atmosphere as the fireball spread out. This infall of material occurred at the same high speeds—Mark Boslough and David Crawford called it a "hypervelocity splat"—and was at least 1000 times brighter than the first precursor. The infall was followed by a lot of surface skittering as material slid horizontally in the stratosphere.

WHAT MATERIALS WERE FOUND AFTER THE IMPACTS?

Water, the most dramatic detection, was found by five teams, including Galileo's NIMS detector, the Anglo-Australian Telescope, two teams on the Kuiper Airborne Observatory, and a team from Italy who noted water on the E site two days after that impact. The 100,000 metric tons of water that Don Hunten and Ann Sprague observed from the Kuiper Airborne Observatory was probably from the comet. Vikki Meadows, observing from Australia's Siding Spring Observatory, also detected water in the K impact, as well as carbon in two forms—first as part of CH_4 (methane), followed by a sudden switch to CO (carbon monoxide)—at a very hot 2000 degrees Kelvin.

Sulfur (S_2) was detected by HST during the G impact on July 18. The telescope detected the sulfur in steadily decreasing quantities until August 9. Two compounds of sulfur, carbon disulfide (CS_2) and hydrogen sulfide (H_2S) were also detected through Hubble's faint object spectrograph by a team headed by Roger Yelle and Melissa McGrath. Other compounds, like ammonia (NH_3) and the hydrocarbon ethylene, turned up.

What are the implications of these chemicals detected after the impacts? Had they come from the planet, they could have provided benchmarks for how deeply the fragments penetrated. But since these materials could have been cometary—Melissa McGrath claims that the metals definitely were—(an astronomer's definition of metals is anything not H or He)—they do not help pin down whether the comets penetrated to Jupiter's layer of water. One could speculate, from the detections of S_2, CS_2, OCS, CO, and HCN (hydrogen cyanide) that the comets penetrated to a level just above the water layer.

WHY WERE THE IMPACT SITES SO DARK?

There is no simple answer to why the impact sites showed up as black eyes on Jupiter. The darkest part of the young spots, the cores, were extremely dark—by far the darkest features ever seen on Jupiter except for shadows of satellites crossing the planet. As Robert West of Jet Propulsion Lab and Gene Shoemaker separately claimed, these cores had considerable depth. The spots were likely mixtures of ices with a collection of organic materials. However, at

Baltimore there were several unanswered questions. What contributed the most to these particles, the comet or Jupiter? Apparently the brown gunk was and is a complicated collection of aerosols that were formed in the core of each impact cloud. As the aerosols settled out, the grains grew in size. Clifford Matthews of the University of Chicago suggested that the dark matter is poly HCN, a hydrogen cyanide polymer, which might have reacted with water ice to form a complex mixture of organic compounds. The dense inner core clouds at the centers of the spots, Gene Shoemaker adds, could also contain sulfur-bearing compounds.

As time passed, the clouds evolved, and much of what we saw, said Reta Beebe of New Mexico State University, was influenced by the local cloud systems. The Q2 spot, for instance, shifted westward with time, and fragment G went right into a cyclonic wind flow region and so was torn apart rather quickly. Since fragment H did not fall into such a constrained area, its cloud expanded in all directions. And within a day of its crash, the L site actually took on the shape of the letter "L."

One exciting effect was the ring structure that was so clearly seen in A, E, G, Q1, and R. The rings were apparently an expression of some substance—possibly a hydrogen cyanide polymer—that condensed briefly as a wave passed by. The waves' speed of propagation was constant in time and the same for all these impacts. Waves were the domain of Andrew Ingersoll of Caltech, who considered and eliminated several types of waves in the process of trying to explain the expanding rings, including surface gravity waves (like an ocean wave in deep water), a seismic wave (like those from earthquakes), and two types of sound waves (the speed was too slow for them).

Finally Ingersoll settled on a tropospheric gravity wave. Such a wave is like a wave on the ocean surface, and it propagates both horizontally and vertically. If this wave were trapped in the planet's water layer, and if the ratio of oxygen to hydrogen were some 10 times that of the Sun, then you'd get a wave that behaves exactly as this one does.

Ingersoll's theory has two major implications. One is that

these waves were deep, which implies that the impacting comets went down more deeply than Kevin Zahnle and others calculated. Moreover, this evidence points to the fact that Jupiter's ratio of oxygen to hydrogen is some 10 times that of the Sun. "Ten times more water is needed," Ingersoll concluded, "than if you were building Jupiter strictly out of solar material. If this turns out to be true, that would mean that conditions in the primordial nebula, at the birth of the solar system, would have been different in the region of the outer planets from what they were at the center. It would mean that there was a lot more ice in this region than there was near the nebula's center where the Sun formed.

THE FIZZLES

Two days before the first impact, Paul Weissman's article, "Comet Shoemaker–Levy 9: The Big Fizzle is Coming," appeared in *Nature*.[5] The day after impact A, Weissman spoke over the phone to his mother, who excitedly told him that his name was on the front page of the *Cleveland Plain Dealer*. "What does it say?" he asked. "It says you were wrong!" she said. Over the following months Paul took his well-publicized prediction with humor and aplomb. "Actually," he once told me, "I was half right about predicting fizzles!" True, virtually half the impactors were fizzles. Weissman even offered a "top 10 list" (à la David Letterman) as to why his fizzle prediction was wrong! Among his "reasons":

- I wrote it in Flagstaff at 7000 feet altitude, and my brain was oxygen starved.
- I reviewed it in Pasadena at sea level, and my brain was smog poisoned.
- I was so worried about O. J. Simpson that I wasn't thinking straight.
- What fizzle prediction? I never wrote any prediction for *Nature*! Because I forgot that a ton of feathers and a ton of rocks going 60 kilometers per second each weigh the same. And, finally,

- Everyone knows that fizzle is a Yiddish word meaning, "Great big humongous Jupiter-shaking comet explosion."

At the "fizzles" workshop in Baltimore, Phil Nicholson outlined the classes to which he assigned each of the impacts. Class I included G, K, and L, the largest and most dramatic events. Class II, which included the substantial hits E, H, Q1, R, S, and W, displayed all the major features of class I but at a somewhat reduced level. Note that the brighter of the Q fragments was originally thought to be the biggest piece, since at discovery it was the brightest. The reason the fragment was so bright at discovery was that it was in the process of splitting apart. Class III consisted of the moderate hits A, C, and D.

Seven impacts belonged to the two fizzle classes, IV and V. Class IV included two bright fragments, B and V, which were off the straight line that, Sekanina maintains, included all the original pieces that grew out of the breakup on July 7, 1992. These pieces were bright, not because they were large but because they consisted mostly of smaller particles that were highly reflective. But tiny N also produced sufficient fireworks to be in this class too. Class V impacts, which included F, P2, T, and U, were barely detectable.

THE HUMAN LEGACY OF SHOEMAKER–LEVY 9

Even as the scientists met in Baltimore, people from around the world were recalling their own experiences from the summer of 1994. The biggest impact from Comet Shoemaker–Levy 9 was felt not on Jupiter, but right here on Earth. We saw that in strange ways, from the expression of pure delight on a child's face to the many editorials that raved about the impacts. This was not about science, but about the excitement generated by the impacts and the passion of the astronomers, both professional and amateur, who observed them. Both Shoemaker–Levy 9 and the scientists who watched it came home that week: the comet to its final home on Jupiter, and the scientists to a world that had awakened to the wonder of science. "I regret," noted amateur astronomer Joe Bergeron on CompuServe, "that my local paper interviewed me in the preimpact period and I blithely dismissed S–L 9 as something that would probably be invisible to anyone using less than the Hubble. Instead it turned out to be the most stunning thing I've ever seen in a telescope short of a total solar eclipse."[6]

Larry Lebofsky, a planetary scientist who specialized in asteroids, spent impact week observing at the Richmond Astronomical Society's 31-inch Newtonian reflector. His purpose was not research: He planned to leave with video images of Jupiter suitable for presentation, not to fellow scientists but to children, the harshest judges of all.

Lebofsky has had an impressive career, with more than 75 published papers in scientific journals and a large series of research grants. His major discovery was the existence of water, its molecules locked in clay minerals, on certain asteroids. But in recent years Lebofsky's career goals have changed as he spent less time on research and more time trying to raise children's awareness of astronomy.

Lebofsky knew that the crash of Comet Shoemaker–Levy 9 on Jupiter had the potential to attract wide interest among young people. To take advantage of this, he planned to merge his video images, taken each night during crash week, to create a living, breathing picture of how Jupiter behaved during this week. I helped him get time on the large 31-inch Newtonian reflector telescope that belonged to the Richmond Astronomical Society. With its active members Warren Walker, Shannon Mishey, and David Hartsel, and with his graduate student Andrew Rivkin, Larry found kindred spirit. Their large telescope, with its industrial-strength "might-ee-lift" that lifted groups of people up the 15 or 20 feet to the eyepiece, was almost exclusively dedicated to public observing.

For the first two nights, clouds and a lack of visual reports concerned him. But with the first clear night came a spectacular view of Jupiter, and by the end of the week Larry had obtained some very nice series of images. "Larry Lebofsky," wrote in the

Mansfield *News Journal* in its July 19 editorial, "chose to watch the comet collision here with the amateur astromers at the Warren Rupp Observatory.... We should also appreciate the observatory we have right here in Richland County. It is a treasure that can broaden our horizons."[7]

The first night of the impacts saw star parties all over the world. Some of these were small and impromptu. Linda Pfeiffer, a science teacher from Las Vegas, was camping in Utah with friends, when she heard the park ranger say he'd set up a telescope near the visitor center for Jupiter observations. Thirty campers got a look at Jupiter. High on Mount Wilson, near Los Angeles, hundreds of people looked at Jupiter through large professional telescopes and smaller amateur instruments as portable radios squawked news of the mighty plumes seen from Spain and through the Hubble. Jean Mueller, who had helped measure the comet's discovery positions 16 months earlier, was on that mountaintop, along with veteran comet observer Charles Morris.

Inspired by the excitement from the press conference in Washington D.C. on July 16, Russell Sipe, an Orange County amateur astronomer and a regular on CompuServe's online service, set up his 13-inch-diameter Dobsonian telescope. "Wouldn't it be wild, I thought, if I could see other spots on which I had not yet heard reports? When I looked in the eyepiece for the first time that evening and focused on Jupiter there was no sign of a spot. However, there seemed to be a very slight irregularity at the southern edge of the preceding limb. I watched it for a couple of minutes, thinking that it might be an atmospheric effect. After 10 minutes the irregularity became a distinct notch in the limb. Getting excited, I rushed in the house to roust my wife and daughter to take a look. 'Can you see it? Can you see that little notch on the edge of the planet?' They could see it! A few minutes later the notch became a round spot. Then it separated from the limb as it started its march around the planet. Wow! What a sight! I excitedly told my 10-year-old daughter and my wife that what they were witnessing was an event unlike any mankind had ever seen. This was celestial history in the making right before our very eyes. I too wanted to rush into a room like Heidi Hammel, champagne in hand and say 'I can see it!' July 17, 1994, will be etched in my memory for all time."[8]

Sipe and many others communicated over CompuServe and other online services. Soon after impact week CompuServe was offering more than 750 impact images to its subscribers. By early 1995, users had downloaded more than 11.5 *billion* bytes of S–L 9 related information.[9]

Clark Chapman's public education event was completely unintentional, but highly effective. Although Clark, a scientist for the Galileo spacecraft, had some time during impact week to look at the big spots from G and L, his schedule for the end of the week was hectic and tense. He planned to travel to Prague for an auspicious meeting of the Meteoritical Society on impacts, but at the same time, using the laptop computer he brought with him, he intended to work by e-mail with other members of the Galileo imaging team to decide which impact data to bring back from the spacecraft. Time was critical, for the spacecraft needed to stop beaming back its impact data by the end of January, 1995, a tall order for Galileo's small antenna.

So with his wife Lynda, Clark flew to Munich and continued by train to Prague. Once off the train, they searched in vain for the subway that they had been told was nearby. Clark left Lynda for the briefest of moments—it had to be less than 50 seconds. When he returned, Lynda was there, but his carryon bag—with his computer—was gone.

Lynda had been "two-teamed," it turned out, by a pair of thieves. One distracted her while the other grabbed Clark's bag. "I absolutely needed that computer for S–L 9 business!" Clark complained to the police, who had a full-time translator just for tourists who'd been robbed. The translator did not act as though she cared that Clark had only a few days to work with his fellow Galileo mission scientists to decide which images to bring down from the distant spacecraft. To add insult to injury, his call to Tucson to report the theft set him back 100 dollars.

The organizers of the meeting tried to help by putting a notice about the theft in the local newspaper. Meanwhile, Clark delivered a talk at the meeting about Shoemaker–Levy 9. "I emerged after the talk to find television cameras, reporters with microphones, people with translators—in the hallway." They asked about the theft, and wondered what sort of genius Clark was, a scientist who could give his S–L 9 talk with everything stolen. Clark's picture was on the front page of a local paper, as was the story of the missing S–L 9 computer.

The next day, more than one local newspaper blasted the Prague police for ignoring Clark's complaint, and Clark's talk on S–L 9, given in fine professional form despite the loss of his computer, elevated him to the status of a hero. At a reception that afternoon, as Clark listened to the mayor's address, translated from Czech, he noticed that His Honor was looking straight at him. "I am a scientist too," the mayor was explaining, "and I remember having my slides stolen before a talk." Both the mayor and the vice-mayor, who happened to be a geophysicist, then came over to Clark to commiserate and talk about comet impacts.

Clark had traded his computer for a whole city's personal involvement with the impacts. While the people of other cities were watching the battered planet in space, the citizens of Prague got an unexpected look at events which gave them an insight on how scientists work, and live, and how two petty thieves could bring a Jupiter scientist quickly back to Earth.

HOW LUCKY WERE WE?

In his final summary talk at the Baltimore conference, Gene Shoemaker left his audience of scientists with an impression of how rare an event like Shoemaker–Levy 9 actually is. The frequencies he described were for a comet 1.5 kilometers in diameter; if the progenitor comet were larger than that, then the event we saw would be even scarcer.

A 1.5 kilometer comet should strike Jupiter every century or so. Such a comet hitting Jupiter should have left a massive dark spot, probably larger than any of the S–L 9 spots. But statistics are figured over the long term, and S–L 9 could well have been the first major impact in a long time. Only half that often—once every 200 years—should a comet strike Jupiter while orbiting the planet.

A comet's breaking up while in orbit and then hitting Jupiter on the next pass would be far less frequent. If the breakup happened two or more orbits before impact, the individual pieces would long since have gone off on separate orbits, with some hitting Jupiter at widely separated intervals and others escaping. So the ideal situation is for the comet to break up on one pass, releasing great quantities of dust that brighten it up so that it can be discovered and studied. Then the comet fragments hit on their next pass, while they still have similar orbits. The most optimistic frequency for such an event is once every *2,000 years.*

"Now what are the odds," Gene concluded, "to have such a rare event happen in the decade that all the new infrared detectors became available, as the Galileo spacecraft was in position to see the hits directly, and only six months after the Hubble was fully operational—AND before the expected cutbacks make the money run out?"

"Folks," he exclaimed, "we had a bloody miracle."

As he spoke those words, 700 miles to the northeast the Fundy tides were rushing out, continuing their twice-a-day demonstration of the gravitational forces at work in the solar system. Almost 500 million miles away, Jupiter still bore a belt of dark material—all that is left of Comet Shoemaker–Levy 9 at the end of its gravitational dance with the giant planet. Comet crashes and life have gone hand in hand in the long history of Earth, from the early times as life was gaining a foothold, to the extinction of the dinosaurs 65 million years ago. In its destruction on Jupiter, Comet Shoemaker–Levy 9 gave us a precious reminder of our heritage.

References

CHAPTER 1

1. C. Chapman, *Sky and Telescope* **35** (1968), 276–279.

CHAPTER 2

1. D. Levy, *The Quest for Comets: An Explosive Trail of Beauty and Danger* (New York: Plenum, 1994).
2. E. M. Shoemaker, personal communication to B. G. Marsden, March 26, 1993.
3. B. G. Marsden, personal communication, March 25, 1993.
4. J. V. Scotti, personal communication to B. G. Marsden, March 26, 1993.
5. *Arizona Daily Star*, March 1993.
6. D. Jewitt, personal communication to B. G. Marsden, March 27, 1993.
7. C. W. Tombaugh, personal communication, 1985. See also D. H. Levy, *Clyde Tombaugh: Discoverer of Planet Pluto* (Tucson: University of Arizona Press, 1991), 61.
8. M. Lindgren, "Jupiter as the Arbiter of Comets" (Sweden: Löfgren and Öberg Tryck AB, 1995), Paper II, 7.

9. *Southern Sky*, May/June 1994.
10. O. Naranjo, personal communication to B. G. Marsden, March 30, 1993. See also *IAU Circular* 5744, April 3, 1993.
11. O. Naranjo, personal communication to D. Levy, September 10, 1994.
12. B. C. Willard, *Russell W. Porter: Arctic Explorer, Artist, Telescope Maker* (Freeport, Maine: Bond Wheelwright, 1976), 145.
13. A. Ingalls, "The Heavens Declare the Glory of God," *Scientific American*, November, 1925, 293–95.
14. Willard, 187.
15. The description comes from his student, James Edson, personal communication, August, 1992.

CHAPTER 3

1. J. Mellosh and P. Schenk, *Nature* **365** (1993), 731.
2. B. G. Marsden, "Original and Future Cometary Orbits IV," *Astronomical Journal* **99** (1990), 1971–1973.
3. Seneca, *Quaestiones Naturales*, VII, "De Cometis" XVI, 2.
4. B. A. Smith, "Voyager 2 at Saturn," *Science* **212** (1981), ring discussion, 182–190.
5. B. G. Marsden, *IAU Circular* 5744, April 3, 1993.
6. *Minor Planet Circular* No. 21987, May 6, 1993.
7. B. G. Marsden, personal communication, February 4, 1995.
8. *IAU Circular* 5800, May 22, 1994.

CHAPTER 4

1. *Time*, May 23, 1994, 59.
2. B. G. Marsden, personal communication, August 28, 1993.
3. G. Kronk, *Comets: A Descriptive Catalog* (Hillside, N.J: Enslow, 1984), 5.
4. *Ibid.*, 68.
5. *Ibid.*, 222–24.
6. B. G. Marsden, personal communication, February 4, 1995.
7. Kronk, 189.

CHAPTER 5

1. A. P. Ingersoll's "Jupiter and Saturn," offers a good technical summary of our knowledge of Jupiter. It appears in J. K. Beatty, B. O'Leary, and A. Chaikin, *The New Solar System*, 2nd ed. (Cambridge, Massachusetts: Sky Publishing Corporation, 1982), 117–129.
2. B. A. Smith, "The Voyager Encounters," in Beatty *et al.*, 109.
3. Ingersoll, 122.
4. *Ibid.*, 124.
5. J. W. Van Allen, "Magnetospheres and the Interplanetary Medium," in Beatty *et al.*, 26.

CHAPTER 6

1. C. Chapman, "Comet on Target for Jupiter," *Nature* **363** (1993), 493.
2. B. G. Marsden, personal communication, February 4, 1995.
3. B. G. Marsden, personal communication, March 30, 1994.
4. L. C. Peltier, *Starlight Nights: The Adventures of a Star Gazer* (New York: Harper & Row, 1965), 2.
5. *London Times*, advertisement, February 8, 1994.

CHAPTER 7

1. L. MacFadden and M. A'Hearn, e-mail announcement, November 16, 1993.
2. Z. Sekanina, Baltimore, January 10, 1994.
3. S. Larson, Baltimore, January 10, 1994.

CHAPTER 8

1. In his *Cosmic Impact* [(London: Fourth Estate, 1986), 169], J. K. Davies suggested that the plight of Damocles was similar to the threat we face.
2. C. Chapman, "Hazard to Civilization of Asteroid and Cometary Impacts," Asteroid Hazard Conference, USSR Academy of Sciences, St. Petersburg, October 10, 1991.

3. B. G. Marsden, "The Next Return of the Comet of the Perseid Meteors," *Astronomical Journal* **78** (1973), 658.

4. *Ibid.*, 656.

5. *IAU Circular* 5636, October 15, 1992.

6. B. G. Marsden, personal communication, October 12, 1992.

7. *Newsweek*, October 1992.

8. B. G. Marsden, interview, January 7, 1993; personal communication, February 4, 1995.

9. D. Lindley, "Earth Saved from Disaster!" *Nature* **360** (1992), 623.

CHAPTER 9

1. Z. Sekanina, personal communication, March 17, 1994.

2. H. Weaver, personal communication, March 10, 1994.

3. P. Weissman, "Comet Shoemaker–Levy 9: The Big Fizzle Is Coming," *Nature* **370** (1994), 94–95.

4. E. Asphaug and W. Benz, *Nature* **370** (1994), 120–124.

5. Weissman, 95.

6. B. G. Marsden, personal communication, March 30, 1994.

7. H. Weaver, personal communication to B. G. Marsden, July 16, 1994.

8. J. V. Scotti, personal communication, July 16, 1994.

9. R. Marcialis, personal communication, July 15, 1994.

10. *New York Times*, editorial, July 16, 1994, p. 20.

CHAPTER 10

1. T. S. Eliot, *The Hollow Men*, in *Collected Poems 1909–1962* (London: Faber and Faber, 1974), 92.

2. H. A. Weaver, personal communication to B. G. Marsden, July 16, 1994.

3. H. A. Weaver, personal communication, February 23, 1995.

4. H. A. Weaver, personal communication, February 23, 1995.

5. "The Observing Campaign Begins," press conference, Space Telescope Science Institute, Baltimore, July 16, 1994.

6. T. Herbst *et al.*, SL9 Message Center, University of Maryland, July 16, 1994. MAGIC is the Max Planck Institute's infrared camera.

7. R. M. West, European Southern Observatory; SL9 Message Center, University of Maryland, July 16, 1994.
8. R. M. West, "Comet Shoemaker–Levy 9 Collides with Jupiter," *ESO Messenger*, September 1994.
9. H. A. Weaver, personal communication, February 23, 1995.
10. H. A. Weaver, personal communication, February 23, 1995.
11. E. M. Shoemaker, Baltimore, July 16, 1994.
12. C. Shoemaker, NASA *Daily Comet Update*, July 16, 1994.
13. *Daily Comet Update*, press conference at Space Telescope Science Institute, July 16, 1994.
14. S. Young, personal communication, November 13, 1994.
15. I. de Pater *et al.*, W.M. Keck Observatory; SL9 Message Center, University of Maryland, July 17, 1994.
16. Z. Sekanina, personal communication, March 17, 1994.
17. S. Miller *et al.*, SL9 Message Center, University of Maryland, July 17, 1994.
18. R. Marcialis, personal communication, July 17, 1994.

CHAPTER 11

1. S. T. Coleridge, *Kubla Khan*, first published 1798. Coleridge took the two grains of opium to control an attack of dysentery.
2. *Daily Comet Update*, press conference at Goddard Space Flight Center, July 17, 1994.
3. *Daily Comet Update*, July 17, 1994.
4. H. A. Weaver, SL9 Message Center, University of Maryland, July 17, 1994.
5. M. Liu, personal log, October 6, 1994.
6. P. McGregor, SL9 Message Center, July 17, 1994.
7. M. Brown, SL9 Message Center, University of Maryland, July 17, 1994.
8. P. McGregor, SL9 Message Center, University of Maryland, July 17, 1994.
9. R. Baalke, personal communication, November 8, 1994.
10. R. Baalke, personal communication, November 7, 1994.
11. M. Hereld *et al.*, University of Chicago; SL9 Message Center, University of Maryland, July 17, 1994.
12. J. Menzies, SL9 Message Center, University of Maryland, July 17, 1994.

13. T. Herbst *et al.*, SL9 Message Center, University of Maryland, July 17, 1994.

14. R. West, SL9 Message Center, University of Maryland, July 17, 1994.

15. J. Spencer *et al.*, Lowell Observatory, SL9 Message Center, University of Maryland, July 17, 1994.

16. McDonald Observatory Science Team, University of Texas; SL9 Message Center, University of Maryland, July 17, 1994.

17. J. Spencer *et al.*, Lowell Observatory; SL9 Message Center, University of Maryland, July 17, 1994.

18. J. Rogers, SL9 Message Center, University of Maryland, July 17, 1994.

19. A. Shemi and E. Ofek, Givatayim Observatory, SL9 Message Center, University of Maryland, July 18, 1994.

20. P. McGregor, Australian National University; SL9 Message Center, University of Maryland, July 18, 1994.

21. M. Hereld *et al.*, University of Chicago; SL9 Message Center, University of Maryland, July 18, 1994.

22. J. Watanabe, Okayama Astrophysical Observatory, SL9 Message Center, University of Maryland, July 18, 1994.

23. Comet-Jupiter Impact Team, Beijing Astronomical Observatory; SL9 Message Center, University of Maryland, July 18, 1994.

24. Korea Astronomy Observatory Science Team, Kyunghee University Observing Team; S–L 9 exploder, University of Maryland, July 18, 1994.

25. I. de Pater, W.M. Keck Observatory; SL9 Message Center, University of Maryland, July 18, 1994.

26. NASA IRTF Comet Collision Team; SL9 Message Center, University of Maryland, July 18, 1994.

27. J. J. Klavetter, personal communication, October 10, 1994.

28. M. Liu, observing log, July 1994.

29. M. Liu, observing log, July 1994.

30. G. Bjoraker *et al.*, NASA Goddard Space Flight Center; SL9 Message Center, University of Maryland, July 19, 1994.

31. S. Edberg, personal communication, June 22, 1995.

32. R. Marcialis, personal communication, July 18, 1994.

33. T. Hockey, University of Northern Iowa; SL9 Message Center, University of Maryland, July 22, 1994. See also *A Historical Interpretation of the Study of the Visible Cloud Morphology on the Planet Jupiter: 1610–1878.* Las Cruces: New Mexico State University, 1988 (doctoral dissertation).

34. S. O'Meara, personal communication, August 15, 1994.

35. T. Hockey, July 22, 1994.

CHAPTER 12

1. H. Hammel, *Daily Comet Update*, Special Briefing on Fragment G, July 18, 1994.
2. C. Sagan, Planetary Society Lecture, July 22, 1994.
3. P. McGregor and M. Allen, Australian National University; SL9 Message Center, University of Maryland, July 19, 1994.
4. R. Marcialis, personal communication, July 19, 1994.
5. *Daily Comet Update*, July 20, 1994.
6. A. Cochran, *Daily Comet Update*, July 20, 1994.
7. M. Liu, July 20, 1994.
8. J. Bell *et al.*, Lick Observatory Crossley 36"; SL9 Message Center, University of Maryland, July 19, 1994.
9. M. Skrutskie and S. Aas, Whately Observatory; SL9 Message Center, University of Maryland, July 19, 1994.
10. Nordic Optical Telescope; SL9 Message Center, University of Maryland, July 19, 1994.
11. M. Skrutskie and S. Aas, SL9 Message Center, University of Maryland, July 19, 1994.
12. I. McGregor, McLaughlin Planetarium, personal communication, November 18, 1994.
13. M. Sykes, University of Arizona; SL9 Message Center, University of Maryland, July 19, 1994.
14. T. Livengood, European Southern Observatory; SL9 Message Center, University of Maryland, July 19, 1994.
15. R. P. Binzel, MIT; SL9 Message Center, University of Maryland, July 19, 1994.
16. J. Rogers, J. Lancashire, and J. Shanklin; SL9 Message Center, University of Maryland, July 19, 1994.
17. D. Jewitt and P. Kalas, University of Hawaii; SL9 Message Center, University of Maryland, July 19, 1994.
18. J. Rogers, Cambridge University; SL9 Message Center, University of Maryland, July 19, 1994.
19. Bill Sandel, Lunar and Planetary Lab; SL9 Message Center, University of Maryland, July 20, 1994.
20. French–Spanish–Swedish observing team at La Palma; SL9 Message Center, University of Maryland, July 20, 1994.

21. T. Herbst *et al.*, Calar Alto Observatory; SL9 Message Center, University of Maryland, July 21, 1994.

22. I. de Pater, Keck Observatory; SL9 Message Center, University of Maryland, July 21, 1994. "We detected the impact from the lost fragment M at around 6:08 UT. The impact site was clearly separate from that of K, which came into view a little later."

23. *Daily Comet Update*, July 21, 1994.

24. *Daily Comet Update*, July 20, 1994.

25. E. M. Shoemaker, interview, February 17, 1993.

26. N. Armstrong, White House Address, July 20, 1994.

CHAPTER 13

1. "Second star to the right, and straight on till morning" are the directions to Neverland. J. M. Barrie, *Peter Pan*, 1911, Chapter 4, "The Flight to Neverland."

2. Michael Skube, "Something New Under the Sun," *Atlanta Constitution*, July 31, 1994.

3. J. DeYoung, United States Naval Observatory, personal communication, November 7, 1994.

4. W. Wild, University of Chicago; personal communication, October 4, 1994.

5. S. Edberg, private communication, June 22, 1994.

6. *Prince George's Journal*, Lanham, Maryland, July 21, 1994.

7. Editorial, *Binghamton Press and Sun-Bulletin*, Binghamton, New York, July 22, 1994.

8. Editorial, *Boston Globe*, July 23, 1994.

9. "Thank Heavens, No O. J.!" Editorial, *Primos Times*, Primos, Pennsylvania, July 19, 1994.

10. Starbeams, *Kansas City Star*, July 19, 1994.

11. "Events in space are inspiring," Editorial, *Elizabeth City Advance*, Elizabeth City, North Carolina, July 25, 1994.

12. *Taunton Gazette*, Taunton, Massachusetts, July 22, 1994.

13. E. Asphaug and W. Benz, "Density of Comet Shoemaker–Levy 9 Deduced by Modeling Breakup of the Parent 'Rubble Pile,'" *Nature* **370** (1994), 120–124.

14. R. Beebe, *Jupiter* (Washington, D. C.: Smithsonian Institution Press, 1994), 226.

15. R. Beebe, *Daily Comet Update*, July 21, 1994.

16. F. Cordova, *Daily Comet Update*, July 21, 1994.

17. Angelo Hall, *An Astronomer's Wife: The Biography of Angeline Hall* (Baltimore: Nunn and Co., 1908), 88.

18. Angelo Hall, Ref. 17.

19. J. Montani, Lunar and Planetary Laboratory, University of Arizona; SL9 Message Center, University of Maryland, July 22, 1994.

20. I. de Pater *et al.*, W.M. Keck Observatory; SL9 Message Center, University of Maryland, July 21, 1994.

21. Crisp, Anglo-Australian Telescope; SL9 Message Center, University of Maryland, July 21, 1994.

22. P. Nicholson *et al.*, Palomar Observatory; SL9 Message Center, University of Maryland, July 21, 1994.

23. The McDonald Comet Impact Science Team; SL9 Message Center, University of Maryland, July 21, 1994.

24. P. Nicholson *et al.*

25. W. Harris, University of Michigan, *Daily Comet Update*, July 22, 1994.

26. W. Harris and the IUE Science Team; SL9 Message Center, University of Maryland, July 21, 1994.

27. *Ibid*.

28. K. S. Noll, Space Telescope Science Institute; SL9 Message Center, University of Maryland, July 21, 1994.

29. G. Rieke *et al.*, Steward Observatory; SL9 Message Center, University of Maryland, July 21, 1994.

30. M. L. Kaiser *et al.*, Goddard Space Flight Center; SL9 Message Center, University of Maryland, July 21, 1994.

31. T. Herbst *et al.*, Calar Alto Observatory; SL9 Message Center, University of Maryland, July 21, 1994.

32. McDonald Observatory Science Team; SL9 Message Center, University of Maryland, July 22, 1994.

33. P. Weissman, Jet Propulsion Laboratory; SL9 Message Center, University of Maryland, July 22, 1994.

34. E. M. Shoemaker, personal communication, November 30, 1994.

CHAPTER 14

1. T. W. Hill, Rice University; SL9 Message Center, University of Maryland, July 22, 1994.

2. A. Ingersoll, *Daily Comet Update*, July 22, 1994.
3. "Hubble's Low-Key Excellence," in *Space News*, July 25, 1994. Courtesy R. Stachnik; SL9 Message Center, University of Maryland.
4. *Daily Comet Update*, July 23, 1994.
5. B. G. Marsden, personal communication, February 7, 1995.
6. The Palomar Caltech–Cornell comet crash team: P. Nicholson *et al.*; SL9 Message Center, University of Maryland, July 25, 1994.
7. M. J. Klein and S. Gulkis, Jet Propulsion Laboratory, California Institute of Technology; SL9 Message Center, University of Maryland, July 28, 1994.
8. E. Ryan, personal communication, June 9, 1995.
9. P. Leonard, SL9 Message Center, University of Maryland, July 29, 1994.

CHAPTER 15

1. H. A. Weaver *et al.*, "The Hubble Space Telescope (HST) Observing Campaign on Comet Shoemaker–Levy 9," *Science* **267** (1995), 1287.
2. *Ibid.*, 1282.
3. H. B. Hammel *et al.*, "HST Imaging of Atmospheric Phenomena Created by the Impact of Comet Shoemaker–Levy 9," *Science* **267** (1995), 1288.
4. H. A. Weaver *et al.*, 1285.
5. P. Weissman, "Comet Shoemaker–Levy 9: The Big Fizzle Is Coming," *Nature* **370** (1994), 94–95.
6. J. Bergeron to S. H. Lucas, CompuServe Astronomy Forum, June 11, 1995.
7. Mansfield *News Journal*, editorial, July 19, 1994.
8. R. Sipe, personal communication, February 16, 1995.
9. D. DeLoach, personal communication, January 20, 1995.

Index